PREFACE

The findings of evolutionary biology are deeply integrated into our culture. They are so deeply integrated that they cannot be excised. Evolution informs agriculture, medicine, public health, environmental health, natural resource management, human understanding, and even the pursuit of justice within the legal system. Scientific understanding of natural phenomena and the social acceptance of that understanding are not necessarily linked, however. Evolution's great value as a successful set of concepts for explaining nature is augmented by unusual public interest, albeit in one very small corner of its biological study area. That interest focuses on the historical relationship between one species, ours, and other life forms, as well as the role of natural selection in that relationship. If, somehow, people were not involved, evolution's light on the cultural radar screen would be relatively dim; perhaps closer to that of photosynthesis or gravity. But people are involved and, although the fact of common descent cannot be undone, some would like to do so. Evolution's role in public education, publicly funded research, scientific literacy, and religious culture is substantial and irresistible in many ways. That is why I have written this book. Greater understanding and interest in the useful applications of evolutionary biology may help in its ongoing assimilation by diverse cultures.

Like so many other biologists, my intellectual journey began along the naturalist's path. The experiences of learning about organisms' variability and life histories, firsthand, have been invaluable. What I found tremendously appealing was, and still is, the potential for novel insight about the near-infinite details of nature. I owe a great deal to my early mentors Frank Craighead and Clayton White and to the many friends and colleagues with whom I have enjoyed discussion and debate. For their comments on all or portions of the draft manuscript, I would particularly like to thank Richard Alexander, Joseph Brown, Dan Graur, Jeff Johnson, Heather Lerner, Diarmaid Ó Foighil, Robert Payne, Joshua Rest, Jack Sites, and John Vandermeer. John Megahan provided the original drawings for page ii, figures in Chapters 2 and 3 and valuable assistance with many others. I thank my family for their encouragement and understanding.

THE EVOLVING WORLD

THE EVOLVING WORLD

Evolution in Everyday Life

DAVID P. MINDELL

HARVARD UNIVERSITY PRESS
Cambridge, Massachusetts
London, England

Printed in the United States of America

First Harvard University paperback edition, 2007

Library of Congress Cataloging-in-Publication Data

Mindell, David P.
The evolving world : evolution in everyday life / David P. Mindell.
p. cm.
Includes bibliographical references (p.).
ISBN-13 978-0-674-02191-4 (cloth: alk. paper)
ISBN-10 0-674-02191-6 (cloth: alk. paper)
ISBN-13 978-0-674-02558-5 (pbk.)
ISBN-10 0-674-02558-x (pbk.)
1. Evolution (Biology). 2. Evolution (Biology)—Social aspects. I. Title.

QH371.M54 2006
576.8—dc22 2005058131

CONTENTS

THE EVOLVING WORLD

If a man will begin in certainties he shall end in doubts; but if he will be content to begin in doubts he shall end in certainties.
—Francis Bacon (1620)

1

A BRIEF HISTORY OF THREE UNPOPULAR DISCOVERIES

All 1,250 copies of Charles Darwin's *Origin of Species* sold out on the first day after publication. His meticulously supported ideas were astounding to the educated lay reader and displeasing to many within the religious establishment. The brilliance and power of Darwin's work lay in his ability to marshal disparate existing pieces of evidence and show how descent with modification, via natural selection, could help explain them all. Over time, both the foundations and the ensuing structure of evolutionary theory have grown prodigiously as new discoveries support and extend Darwin's conclusions. Despite overwhelming evidence supporting evolution and more than a century of acceptance within the scientific community, however, acceptance of evolution within some communities, particularly fundamentalist religious ones, still lags.

But Darwin's theory is hardly the only theory in the history of science that has been slow to gain popular approval. In this chapter I discuss two other unpopular theories—heliocentrism and germ theory—that provide some context for understanding how scientific theories eventually gain acceptance, then I move on to explore the intellectual culture and history leading up to the publication of *Origin of Species.* Lest there be any confusion, I do not mean "theory" in the pejorative sense of "speculation," but do use "theory" in the scientific sense, as including the full set of ideas, facts, and analytical methods relevant to studying natural phenomena.

Heliocentrism

Nicholas Copernicus' treatise *De revolutionibus* was published as he lay dying in 1543. In this scholarly work, he presented evidence for a planetary system, with the sun standing in the center while the planets orbited in perfect circles around it. On the day following release of the treatise, nothing much happened. Life went on. Copernicus had actually finished the manuscript about a decade earlier. The heliocentric idea had been circulating among astronomers for thirty years, and it was presented to the public, in a preface by the publication's overseer, as a mere hypothesis, useful in mapping the relative positions of the planets. Discovery that the burgeoning set of astronomical observations better fit a sun-centered system than an earth-centered system was a hit with some mathematicians and astronomers, but missed the masses entirely. The theory had to become better known and to create a more palpable threat to religious authorities before it would gain its legendary unpopularity. This began as Galileo instigated his Copernican crusade, culminating in his own treatise, *Dialogues,* published in February of 1632. About four months later, in July of 1632, Roman Catholic leaders issued an order forbidding further distribution of Galileo's book and recalling copies already distributed. Following Galileo's trial and imprisonment at home for holding and defending Copernicus' heretical views in 1633, *Dialogues* was put on the Index of Prohibited Books, where it remained until 1835.

Unpopularity of the heliocentric discovery derived largely from its perceived contradiction of religious doctrine and breach of higher authority. Galileo repeatedly professed his faith in God and contributed generously and continuously to his church and his daughter's convent, but that did not outweigh his transgression. He also argued that Holy Scripture could not err, though its interpreters may, and that if the intention of the "sacred scribes" had been to teach people astronomy, they would not have passed over the subject so completely. Galileo was not alone in challenging church dogma, just one of the most persistent and compelling. Almost a hundred years earlier, during the 1540s, Flemish anatomist Andreas Vesalius staged frequent, popular demonstrations showing that human male and female skeletons had identical numbers of ribs. This was a clear contradiction of the widespread belief that men must have one less rib than women, based on God having created Eve from one of Adam's ribs as described in the book of Genesis.

The sun-centered cosmology remained controversial and unpopular among some scientists as well as churchmen. Copernicus' calculations of astronomical positions were no more accurate and no easier to complete than those of his predecessors. In addition, Copernicus continued to support some erroneous features of the cosmology that his supplanted, including the existence of solid, planet-bearing spheres and a restricted outermost sphere bearing fixed stars. Parts of his theory were accepted by his contemporaries while the radical core was ignored or rejected by most. Galileo provided strong evidence in support of Copernicus, but he also (erroneously) invoked the movement of the tides as evidence of the earth's daily rotation on its axis and annual orbit about the sun. He reasoned that the Mediterranean Sea, if held in a fixed basin on a stationary earth, could not ebb and flow as it does. Galileo's explanation for tides was uninformed by gravity and thus mistaken. He considered and rejected a possible influence of the moon on the tides because he did not see how a body so far away could exert such force. In 1588, a middle position was developed by the Danish astronomer Tycho Brahe in which the earth remained at rest and all the planets revolved around the sun as it revolved around the earth. After suppression of Copernican theory and censure of Galileo in 1633, many astronomers adopted Brahe's view. With the publication of Isaac Newton's system of celestial mechanics in 1687, most astronomers in England, France, the Netherlands, and Denmark became Copernicans by 1700. Natural philosophers in the other European countries, however, held strong anti-Copernican views for at least another century.

In the end, political pressure by the Roman clergy had no impact on the planets. A sun-centered arrangement became recognized as fact slowly, based on the accumulating weight of evidence, the shortcomings of alternative hypotheses, and an increasing willingness to seek natural rather than supernatural explanations by a growing segment of the population. Though self-correcting in the long term, science is prone to politics and personality clashes in the short term. Further, simple, decisive experiments are often difficult to devise, even for science's most talented practitioners. Public opinion is erratic, even without the influence of authoritarian edict, and abstract theorems take a back seat to other factors, ranging from firsthand experience to tradition and superstition. If science provides a useful application for a theory, acceptance of the underlying principles is more likely. In the case of heliocentrism, no such immediate application existed. Accurate navigation of the world's oceans relied on the later development of instruments to measure time and the relative position of the sun or a star, rather than an understanding of planetary movements over longer time periods.

Human explanations for the physical universe appear to be as old as our capacity for language. Chinese mythology holds that sky and earth, initially undifferentiated, separated during an 18,000-year interval in which the yang, being light and pure, rose to become the sky and yin, which was heavy and turgid, sank to form the earth. North American Cherokee legend holds that the earth was an island surrounded by water and suspended from the sky by cords of rock. The works of Copernicus, Galileo, and other astronomers were similarly motivated by a desire to understand and explain the natural world. However, their work marks a significant shift toward natural rather than supernatural approaches to understanding cosmology. Combined with rigorous methods of testing, natural approaches proved to be both resilient in the face of repression and intellectually compelling across cultures and over time, as my next example suggests.

Germ Theory

Understanding the arrangement of planets and the causes of disease, though different in detail, share a long journey from supernatural to natural explanation. For thousands of years, human disease and epidemics were viewed as divine judgments on the wickedness of humans. In the mythology of ancient Egypt, fearsome Sekhment, with a lion's head on a woman's body, reigned as the goddess of pestilence. She started plagues when she was angry and stopped them when appeased. Zeus, chief of the Olympian gods, created Pandora in order to punish mankind for their sins, and when Pandora removed the lid from the jar given to her by Zeus, evil and disease flew out to scourge the mortals. In Homer's *Iliad,* written about 800 BCE, gods induced plagues at will to assuage their needs for vengeance, and in Hebrew scriptures and lore over 5,000 years old, the leprosy of Moses' sister Miriam, the boils of long-suffering Job, and dysentery of Jehoram, an Israelite king, were all attributed variously to the wrath of God or malice of Satan.

Regardless of causation, there has also been a long history of practical care for the sick and injured. One of the best-known early records is the papyrus acquired in Luxor in 1862 by Edwin Smith. This ancient Egyptian medical treatise, believed to be a copy of a work dating from about 3000 BCE, describes the diagnosis and treatment for a variety of injuries. Brief instruction is given for counting the pulse as a measure of heart activity and for closing of wounds with sutures, followed by application to the wound surface of raw meat or grease and honey. Descriptions are also provided of physically resetting dislocated shoulders and jaws and for treating a broken nose. The latter seems timeless, as an ancient author recommends carefully cleaning the blood from the nostrils, returning the nasal bone to the normal position, packing the nostrils with linen saturated with grease, and applying stiff linen splints on either side of the nose. The association between head and neck injury and paralysis of limbs is clearly drawn. Though there is no discussion of infectious disease, we can clearly recognize aspects of a modern scientific approach, particularly careful observation and the lack of supernatural theorizing.

Early association between the spread of disease and contact with the sick, rancid food, or water contaminated with feces led to institution of both personal and group hygiene practices in some populations. Although such practices did not entail scientific understanding of causes of disease, they were certain to provide some benefit. Evidence for sanitary practices dates back at least 4,000 years to former cities in northern India, where bathrooms with drains were common and under the paved streets lay a covered sewer system lined with mud bricks.[1] From about the same time, the Old Testament refers to many rituals or prohibitions regarding social and personal hygiene. By one accounting, of 613 commandments in Jewish scriptures, 213 are of a medical nature designed to promote the health and well-being of the community.[2] For example, to keep military camps clean, soldiers were given explicit instructions for construction of latrines and covering of waste (Deuteronomy 23: 13–15). Those with apparently infectious diseases, including leprosy, were excluded from the camp for specific quarantine periods (Leviticus 15:1–15; Numbers 5:1–4). The Babylonian Talmud, completed in about 500 CE but summarizing millennia of oral tradition, states that no tannery, grave, or carcasses may be placed within 200 feet or so of a human dwelling, and that streets and market areas be kept clean. Scholars were urged not to reside in any city without a public bath.

Early Theories of Disease Origins

The beginning of a scientific theory of disease is traced to the Greeks, and particularly Hippocrates, born about 460 BCE, who laid the foundations of medical science on natural causes and observation. He emphasized the influence of diet, occupation, and especially climate and the environment in causing disease. Epilepsy was widely known at the time as "the sacred disease," and, indicating his approach, Hippocrates wrote, "It is not any more sacred than other diseases, but has a natural cause, and its supposed divine origin is due to man's inexperience. Every disease has its own nature, and arises from external causes." The association between malaria and habitation near swamps was well known during his time, though it was mistakenly thought that malarial fever resulted from drinking water or breathing "bad air" from the marshes. Understanding of the role of the blood-borne, protozoan parasites in the genus *Plasmodium* carried by mosquitoes would be a long time coming. Though Hippocrates' views were based on natural rather than supernatural causes, many of his theories regarding human attributes, among other things, are thoroughly mistaken. In his treatise *On Airs, Waters, and Places,* he says, "the principal reason the Asiatics are more unwarlike and of gentler disposition than the Europeans is the nature of the seasons, which do not undergo any great changes either to heat or cold." As a pioneering physician, Hippocrates had abundant opportunities to record the first mistaken medical views. Yet nestled among his flawed analyses lies a good deal of common sense. The Hippocratic Oath is still often cited at ceremonies where medical degrees are conferred. It includes the injunction to avoid causing harm and to prescribe only beneficial treatments—advice that has stood the test of time.

Sextus Julius Frontinus (c. 40–104 CE), a Roman soldier, governor, and superintendent of aqueducts, demonstrates clear understanding of the importance of clean water to public health in his book, "Concerning the Waters of Rome." He relates that many water management projects, which began as early as 312 BCE, succeeded in reducing incidence of disease, alleviating numerous problems that had been giving Rome a bad reputation. The Romans brought their engineering and administrative skill to bear on the issue of providing clean water and waste removal to a few segments of society, decreasing the incidence of disease.

Impediments to an Empirical View of Disease

Despite our ancestors' demonstration of the human capability for observation and logic, cultural and religious forces superseded a scientific approach following dissolution of the Roman Empire in the fifth century. Instead of reliance upon observation, experience, and experiment, attention focused on sacred documents and supernatural agencies during the Middle Ages.

A primary impediment to progress was the Christian church's injunction against human anatomical study and dissection. For over a thousand years surgery was considered dishonorable, while miracle and fetish cures were pursued with fervor. The water in which a single hair of a saint had been dipped was prescribed as a purgative, and wine which had been poured over the bones of a saint were said to cure lunacy. Numerous saints gained popularity for curing particular diseases, including St. Valentine for epilepsy, St. Gervase for rheumatism, and St. Apollonia for toothache. Church authorities were scathing in their opposition to medical cures, especially if administered by non-Christians. In the thirteenth century, the free-thinker and twice-excommunicated political foe of the papacy, Frederick II, allowed human corpses to be dissected within regions under his control in present-day northern Italy. In the centuries following, a few other monarchs followed his example, though only timidly. For example, in 1391 John of Aragon granted professors at the University of Lerida the privilege of dissecting one dead criminal every three years.

Though the earliest-known attempts to inoculate against smallpox occurred around 1017 in China, where scabs from smallpox lesions were ground up and inhaled through the nose, inoculation came to Western societies much later. In 1716, Lady Mary Wortley Montagu, wife of the British Ambassador to Turkey, wrote a letter describing the widespread and popular inoculations with smallpox scab powder that took place each September in Constantinople. This practice, already widespread among peasants elsewhere, was reported by travelers in Poland, Denmark, Scotland, and Greece, and soon became common among the aristocracy. In 1721, Zabdiel Boylston, a physician in Boston, made an experimental inoculation. Beginning with his son and two slaves, he inoculated over 240 persons, all but six of whom survived. Public sentiment, however, was against the experiment, and the lives of Boylston and his supporter Cotton Mather, a minister, were threatened. Mather's house was attacked with explosives, forcing them to conduct the work in secrecy. Boylston's opposition insisted that inoculation was no better than poisoning, and they urged the local authorities to try him for murder, despite the much lower mortality rate of his inoculants compared to the rest of the population during the epidemic. His religious opponents insisted that smallpox was a judgment rendered by God on the sins of the people and that "to avert it is . . . an encroachment on the prerogatives of Jehovah, whose right it is to wound and smite." European theologians solemnly condemned it as well. Benjamin Franklin lobbied in favor of inoculation in the 1750s, having lost a son to smallpox in Boston, where as many as 38 percent of the population were infected. But even over fifty years after Boylston's experiments, Reverend Edward Massey published a sermon in England in 1772 speculating that Job's ancient distemper was a result of smallpox induced by inoculation from the devil.

In the 1770s Edward Jenner, an English physician, found that human inoculation with material from cowpox lesions (soon termed "vaccination," after the Latin *vacca* for cow) could also provide immunity. Though he gained popular acclaim for his discovery, he was also attacked. In 1798 an Antivaccination Society was formed in Boston by physicians and clergymen, who called on the masses to suppress vaccination as "bidding defiance to Heaven itself, even to the will of God." Despite these protests, and despite an incomplete understanding of vaccination's mechanism, the success of treatments led to an increase in the number of people inoculated. General George Washington ordered all American soldiers in the Revolutionary War to be inoculated against smallpox, and once vaccines from cowpox became available, President Thomas Jefferson worked to establish vaccination as a public health procedure for the new country.

In the early 1600s Jesuit missionaries in South America learned from the natives the effectiveness of "Peruvian bark" in the treatment of the fever and chills of malaria. Its constituent alkaloid—quinine—succeeded dramatically in reducing death rates wherever it was used. Its use was bitterly opposed by many conservative members of the medical profession and by large numbers of devout Protestants simply out of hostility to the Jesuits. The new remedy was stigmatized by some opponents as "an invention of the devil."

Even knowledge of pain relief was suppressed. In 1591, Eufame Macalyane was charged with seeking aid for the relief of pain during childbirth and was burned alive for this crime on the Castle Hill of Edinburgh, Scotland. As recently as 1847, James Young Simpson, a Scotch physician, was denounced from the pulpit for pioneering the use of chloroform as an anesthetic in difficult cases of childbirth. Holy Writ was cited to support the argument that use of chloroform was an attempt to "avoid one part of the primeval curse on woman." In a clever turnabout, Simpson used the Old Testament in defense of anesthetics, invoking the story of Genesis as a record of the first surgery ever performed, in which God "caused a deep sleep to fall upon Adam" prior to extracting a rib for the creation of Eve.

Despite the proclamation of John Wesley, founder of the Methodist church, that "cleanliness is near akin to godliness," for centuries the opposite idea had prevailed. In the fourth century Saint Hilarion, a healer and exorcist who renounced society, lived in studied uncleanliness. Saint Euphraxia belonged to a convent in which the nuns religiously abstained from bathing. Saint Simon Stylites, who lived for many years in austerity atop a pillar fifteen meters tall, was said to have lived in a stench barely tolerable to even his most dedicated supplicants.

Between 1347 and 1351, about one-quarter to one-half of the

population of Europe died of the Black Death; in some cities, the death toll approached 80 percent. The disease, now known to be spread by fleas transferring bubonic plague bacteria *(Yersinia pestis)* from rats to humans, thrived where both were found together in large numbers. At the time, a commonly accepted mode of transmission was pollution of the air with poisonous vapors. In spite of this somewhat reasonable if fallacious hypothesis, thousands of lepers, Jews, and other minorities were burned alive by those who believed them to be the agents responsible for this disease. In 1665 in London, another bubonic plague killed more than 100,000 people. Sanitary measures were still few and the plague was generally attributed to divine wrath. In North America, when the plague claimed the lives of so many Native Americans, it was attributed by some to the divine purpose of "clearing New England for the heralds of the gospel." There is one sort of historical calamity that, paradoxically, reduced disease: the periodic great fires that swept through cities, clearing and cleaning them. The town council of Edinburgh declared its great fire of 1700 "a fearful rebuke of God"; however, it was also observed that disease and death were greatly diminished after the fire, though the physical cause was not understood.

Progress in Empirical Approaches

Communal and personal hygiene prolonged more lives than any medicine, but the actual reasons why were not understood for centuries. A naturalist's view, instead of a supernaturalist's view, was needed, and new tools and techniques for studying microbes had to be developed in order to establish "germs" or pathogens as the cause for many diseases. Some highlights are given below, though, as usual, the path to understanding was not direct.

Although the Roman scholar Marcus Varro had proposed

"imperceptible particles" as a cause of disease in the first century BCE, that information was largely disregarded until, in the sixteenth century, Girolamo Fracastoro provided the first scientific theory of disease resulting from infection. Girolamo Fracastoro was a friend and colleague of Copernicus at the University of Padua. He shared the great astronomer's interest in the stars, yet still found time for geology, poetry, and medicine. In 1546, just three years after the death of Copernicus, Fracastoro published *On Contagion and Contagious Diseases,* stating that diseases were caused by rapidly multiplying, imperceptible particles (*seminaria* or seeds) that could be passed among humans by direct contact, via contaminated objects such as clothing, or through the air. Though initially well received, his ideas had little practical impact. Without observations of the hypothesized seeds of disease, his ideas had no advantage over familiar, vague notions of noxious airs. Although Galileo invented the microscope less than seventy years after Fracastoro's publication (by reducing the barrel length of a telescope and looking through the opposite end), distant planets with their regular movements held more interest for the scientists of the time than the swarming particles in a drop of water.

In the mid-1600s in England, the physician Thomas Sydenham likened disease in an individual to a plant that grows and dies. Though derided for his unorthodox views, he advanced the study of epidemics by focusing on observation and natural treatments. He was one of the first to use iron in treating iron-deficiency anemia and helped popularize quinine in treating malaria.

In 1677, Antony van Leeuwenhoek discovered small "animalcules" using a microscope whose lenses he made himself. It would take another two hundred years before a string of discoveries would link similar microbes to specific diseases. Following 640 human dissections, Giovanni Morgagni published

Figure 1.1 Girolamo Fracastoro (1483–1553) published an early hypothesis that diseases were caused by infectious agents in his 1546 book, *On Contagion and Contagious Diseases.*

a book in 1761 showing that different deadly diseases affected different organs in predictable ways and proposed that disease symptoms resulted from these pathological changes in the organs.

In 1767, Lazzaro Spallanzani debunked the theory that van Leeuwenhoek's animalcules were inanimate "vital atoms" responsible for all physiological activities, as had been postulated by Georges de Buffon and John Needham. In a series of experiments he showed that gravy, when boiled, did not yield these atoms if placed in vials sealed with glass, whereas they were observed if the vials were left open to air circulation. Spallanzani went on to demonstrate conclusively that heat could kill these small organisms.

In 1840, Friedrich Gustav Jacob Henle, a German pathologist, argued that the "material of contagions is . . . endowed with a life of its own, which is, in relation to the diseased body, a parasitic organism." By this time, others had already shown that muscardine, a disease of silkworms, was caused by a fungus and that scabies, an itching disease in people, was the result of infestation with mites.

In 1850, Ignaz Semmelweis deduced that the failure of maternity clinic examiners to wash their hands was responsible for the high incidence of childbed fever. When he forced examiners to wash their hands and instruments between visits to different patients and on return from dissections, the rate of fever among mothers dropped from 18 percent to 1 percent. As a result of this beneficial yet incriminating finding, he was vilified and ultimately fired from his position in Vienna.

In the celebrated Broad Street pump incident of 1854, physician John Snow showed that disease pathogens can be transmitted from human feces to humans through drinking water. In a two-week period, over 500 fatal cases of human cholera occurred within 250 yards of the Broad Street pump in London. Snow convinced city officials to remove the pump handle, which led to a rapid decline in the incidence of cholera in the area. Subsequent examination showed that a cesspool was leaking into the well.

Joseph Lister applied Louis Pasteur's findings to treatment of wounds in 1867. Most physicians already knew that scabs decreased rates of wound infection, and Lister hypothesized that sterilizing the wound first would do even better. He was aware of the treatment of sewage in Carlisle, England, with carbolic acid, which both removed the odor and made it safer for cattle to graze on land fertilized with the sewage. Lister applied carbolic acid directly to patients' wounds, and infection rates went down dramatically. He applied sterile gauze treated with carbolic acid to wounds and reduced rates of infection even more.

In 1875 in St. Petersburg, Russia, Fedor Losch unsuccessfully treated a farmer suffering from severe diarrhea. He observed large numbers of amoebae, which he suspected to have caused the disease, in the man's stools. Losch experimentally transmitted these same amoebae and, hence, the disease to other animals, demonstrating their close association. Direct transmission to humans by ingestion of the amoeba *Entamoeba histolytica* would be demonstrated eventually in 1913.

Building on the work of Casimir-Joseph Davaine and Ferdinand Cohn, Robert Koch, a student of Henle, the pathologist, demonstrated in 1876 that the bacteria *Bacillus anthracis* was the cause of anthrax. Koch grew the bacteria in suitable media on microscope slides and found that reproductive cells (spores) isolated from them could remain viable for years, despite exposure to the air. Anthrax is one of the oldest recorded and most devastating diseases of animals, including humans, and Koch's findings not only identified the causative agent, but also explained the recurrence of disease (as a result of persistent spores) long after direct exposure to sources of contamination. Finally, a hypothesis much like Fracastoro's from three centuries earlier was confirmed. A few years later, Koch formulated criteria for proving that a disease was caused by a particular agent. The agent must be found in every case of the disease, grown in a pure form outside of the body, and shown to cause disease when administered to healthy individuals or cells.

Between 1850 and 1870 the French chemist Louis Pasteur engaged in a dramatic series of discoveries. Using an experimental approach, he disproved the notion of spontaneous generation of bacteria, demonstrated that microorganisms can cause fermentation and disease, and worked to develop vaccines for rabies, anthrax, and chicken cholera. Pasteur was challenged to a public test of his anthrax vaccine by a well-known veterinarian, Monsieur H. Rossignol, and he seized the opportunity. In the test, conducted in 1881 at Pouilly-Le-Fort, a small village south of Paris, twenty-five sheep received the vaccine, twenty-five did not, and all received a dose of virulent anthrax. The test results appeared conclusive; all receiving the vaccine survived and all the control individuals died. The experiment, a first public test of the new science of microbiology, was widely reported in French newspapers as well as the London *Times*. The public was convinced, though a bitter dispute began between Pasteur, Koch, and others regarding the role of chance, the methods of vaccine preparation, and its long-term efficacy. It also seems the role of prior work on a vaccine by another veterinarian was given insufficient credit. Thanks to Pasteur's test, within a decade about 3.5 million sheep and half a million cattle were vaccinated with a mortality rate less than 1 percent.

In 1882 Elie Metchnikoff provided an early observation of cellular immunity, a mechanism by which a body could prevent disease. He put a rose thorn into a starfish larva and watched as cells collected around the thorn and devoured it. This indicated that resistance to infection was at least partly due to host cells attacking foreign matter.

As always, discovery spurred invention and vice versa. In 1883, Carl Zeiss teamed up with physicist Ernst Abbe to design microscopes that extended the power of magnification from 600- to about 1,200-fold, improving access to the microbial world. Charles Chamberland developed a porous, porcelain tube in 1884 that could effectively remove bacteria from water filtered through it.

In 1892 Dmitrii Ivanowski in St. Petersburg and Martinus Beijerinck in Delft independently showed that the cause of tobacco mosaic disease was an agent small enough to pass through Chamberland's bacterial filter. They did this by injecting filtered sap from a diseased plant into a healthy one, successfully inducing disease. Thus, these disease agents were called "filterable." They are now known as viruses.

In 1881 the Cuban physician Carlos Finlay suggested that mosquitoes carried the pathogen causing yellow fever, though at that time most thought the disease was spread by poor hygiene and direct contact with an infected person. Finlay's hypothesis was proven in 1898 by a U.S. military commission, led by Walter Reed, in a series of experiments using army and local civilian volunteers at a U.S. garrison near Havana. Mosquitoes were allowed to take blood from yellow fever patients, then kept through an incubation period of at least twelve days (determination of this necessary incubation time was the key to success of the experiments), after which they were allowed to bite uninfected individuals. These experiments showed Finlay's hypothesis to be correct, though the viral agent would not be identified until much later. His tests were instrumental in demonstrating the need for control of mosquitoes.

Theodore Roosevelt, a U.S. Army colonel before he became president, witnessed the ravages of yellow fever firsthand in Cuba during his time as leader of the Rough Riders. During his presidency in 1901 he was quick to implement exhaustive mosquito control efforts during construction of the Panama Canal. Lakes, swamps, and ditches were drained where possible and oil was spread over many water surfaces to destroy mosquito eggs and larvae. Windows, doors, and porches of houses and all railway cars were screened to reduce incidence of mosquito bites. Housing areas were provided with sewers and clean water. By 1906, yellow fever had been wiped out in the Panamanian canal zone.

We can point to the experimental work of Pasteur, Koch, and their contemporaries as the inescapable proof that many diseases are caused by microbes. For most people, however, connecting disease cause with effect is much less impressive than applying the new knowledge to practical measures for reducing disease incidence. In the case of germ theory, practical applications appeared quickly and their results were spectacular, though this required a focus on populations and public health in addition to individual health. The dramatic increase in average life span in many regions of the world over the past century is a result of direct application of our understanding of the role of microbes in infectious disease. Implementation of programs promoting clean water, sewage removal, protection of foods from contamination, and sterile conditions in hospitals have succeeded to the point that acute, infectious diseases are no longer the primary cause of death in some parts of the world.

Germ theory has resulted in a rich network of facts and new hypotheses that marks a major advance for civilization. It took over 330 years to develop both the empirical methods of science and its practical application to human lives that would overcome millennia of superstition and misunderstanding. As often happens, new understanding reveals new realms of ignorance, and the study of disease causation has been a prime motivating force in our attempts over the past century to understand the complex life histories and evolutionary interactions between microbes and their hosts, from the consequences of bacterial population growth to the molecular mechanisms of viral infection and reproduction.

Evolution

From Supernatural to Natural Causation

The history of ideas explaining the origins of life and of humans resembles the history of ideas regarding planetary movements and origins of disease. Explanations begin in antiquity with a basis in supernatural causes and proceed through a lengthy and messy process of observation, hypothesizing, social conflict, testing, discovery, and, fitfully, cultural change. Understanding life's history is arguably the most contentious scientific issue of all, at least for the public, as it is the most personal.

Two of the earliest accounts of human origins for which records survive are contained in Hebrew scriptures and, subsequently, the Christian Old Testament. These accounts were first developed and passed on by semi-nomadic, Semitic people, living in ancient Mesopotamia and the Arabian Peninsula, whose artifacts date back at least to 3000 BCE. The two accounts have been traced to different sources (sets of papyrus scrolls by different authors) among the many used in compiling Hebrew scriptures between 1000 and 350 BCE. The older account, apparently written between 950 and 800 BCE, is known as the J account ("J" derives from the German spelling of the word used for God—Jahweh). It describes the following series of events. God made the first man, Adam, from dust on preexisting earth, God created the Garden of Eden, God created "every beast of the field and fowl of the air" also from dust, and finally, God made a woman from one of Adam's ribs. The P account ("P" derives from "priestly"), apparently written between 700 and 500 BCE, is more familiar to many, and describes creation events by God over six days. God created light; then the heavens; then plants; then sun, moon, and stars; then aquatic creatures and birds; then, on the sixth day, other animals and man, with man and woman created simultaneously.[3]

The existence of two similar but contradictory creation accounts, maintained with care over millennia, implies an ancient disagreement and an uneasy compromise that is characteristically human. Perhaps believers in the divine origin of scriptures could not rationalize the removal of one or the other account simply to promote consistency. The two differing creation accounts have engendered debate for more than two thousand years, and they appear to have played a role in guiding some, including St. Thomas Aquinas in the thirteenth century, to embrace a decidedly nonliteral interpretation of the Bible. The Qur'an, the sacred text of Islam first recorded in about 650, also includes a creation story, broadly similar to those in Hebrew scriptures.

These particular creation accounts became part of the canon of Western civilization following the Christianization of the Roman Empire. This began with the proselytizing efforts of Paul in the middle of the first century and continued at a slow pace for the next 150 years, during which Christianity had no greater following than many other religions and cults of the time. Christianization of the Roman Empire can be considered nearly complete by 392 when Emperor Theodosius officially closed pagan temples and made Christianity the sole sanctioned religion. As a consequence, several early Western concepts crucial to what would become biology can be traced to the Genesis creation accounts. Primary among these are the ideas that all life forms are the direct and intentional result of sudden creation by God and that these life forms have been essentially unchanged ever since. Where the Church ruled as the dominant cultural force, creation mythology was deeply integrated into the prevailing worldview, and inquiry regarding alternative explanations was actively suppressed.

From the fourth to the second centuries BCE, Greek philosophers, including Plato, Aristotle, and Galen, and their patrons pioneered a rich intellectual culture, seeking to explain the physical world using reason and logic. In his book *On the Heavens,* Aristotle says, "The world as a whole was not generated and cannot be destroyed as some allege, but is unique and eternal, having no beginning or end of its whole life." His was not an evolutionary view of life—he said nothing about origins or change in forms over time. Aristotle argued that observable growth and development of biological organisms were the result of an inner driving force that could be called their nature. For example, he would say it is the nature of an acorn to become an oak tree. Ultimately, Aristotle traced all change and motion in the universe to the natures of things, and it was these natures that were the proper objects of study. This may sound evasive to modern ears, but Aristotle's rationalist approach was a major advance. Aristotle is remembered as the founder of the systematic study of organisms based on careful observations of them in nature. Aristotle's argument that physical entities, including organisms, cannot be understood without detailed, accurate knowledge of how they work helped to set the course for future discovery.

The discoveries, however, would be delayed. It was not the shift of political control from Greece to Rome that led to decline in intellectual life. In fact, Rome's growing upper classes were drawn to Greek achievements in science and the arts and sought to continue them. Many Romans spoke both Latin and Greek. Cicero, who translated Plato's *Timaeus,* and Lucretius, author of *The Nature of Things,* are examples of Roman scholars carrying on in the Greek tradition. In the first century Pliny the Elder's encyclopedic *Natural History* attempted to survey the universe and its natural objects. Although his discussions vary from reasoned to nonsensical, his ambition is remarkable in covering topics as diverse as astronomy, geography, anthropology, mineralogy, botany, and zoology. Two centuries of peaceful Roman rule eventually gave way to civil war and economic disaster after the death of Marcus Aurelius in 180 CE. This was exacerbated by attack on the empire's borders starting about 250 CE. By the end of the fifth century, Rome was ruled by foreigners, and the Greek intellectual tradition faltered.

Lack of progress in the sciences following the decline of Greek and Roman culture around 500 CE until the beginning of the Renaissance about 1450 is often attributed to capitulation of the Roman Empire to Christianity. Church censorship certainly played an important role. In his *New Organon,* published in 1620, Francis Bacon describes the times between antiquity and his own era as "unprosperous" for the sciences: "For neither the Arabians nor the Schoolmen need be mentioned, who in the intermediate times rather crushed the sciences with a multitude of treatises, than increased their weight." In the eighteenth century Voltaire decried the "general decay and degeneracy"[4] that characterized the Middle Ages, as did the Marquis de Condorcet, who remarked, "The triumph of Christianity was the signal for the complete decadence of philosophy and the sciences."[5]

Condemnations and burnings of heretics and books by the Church's agents were chilling, yet the Church was not the only impediment to scholarship. The flourishing of science requires political and economic stability, individual patronage, institutional support, and leisure time. All of these were in short supply during much of the Middle Ages. Numerous wars, often accompanied by epidemic disease and famine, swept through the former Roman Empire in Europe, the Middle East, and northern Africa during the sixth and seventh centuries. By the end of the eighth century, Muslim armies conquered Persia, the Arabian Peninsula, northern Africa, and Spain amid great social turmoil. The wars of the Christian Crusaders seeking conquest of the Holy Land during the eleventh and twelfth centuries drained resources that might have been employed elsewhere. There were sustained attacks on Europe and Islamic territories by Mongol armies from east-central Asia (led by Genghis Khan's grandson, Batu) during the early thirteenth century, and warfare was nearly continuous among Islamic factions in the late medieval period, with ongoing wars between Muslims and Christians during the reconquest of Spain from the mid-eleventh century to the mid-thirteenth century.

A growing language barrier also slowed scientific progress during the early Middle Ages, as increasing use of Latin and Arabic reduced access to Greek scholarship among the literate. Translations were undertaken, but only slowly, and in any case, scholars' time was dominated by religious issues. The Roman theologian Boethius translated some portions of the works of Plato and Aristotle into Latin during the sixth century, and this provided the only Latin source for Aristotle's works for over 600 years.

The earlier intellectual culture of the Greeks and Romans thrived not only because of the political and social stability, but also because of the presence of schools promoting free inquiry, some dating back to the sixth century BCE. A prominent example is Aristotle's Lyceum, founded in 335 BCE in Athens. This school provided a place for lectures and debate as well as collaborative research, attracting students and teachers from all over the Mediterranean region. There is a close association between education and urbanization: urbanization can concentrate wealth and talent, which fosters growth of schools. Given the changing demographics and the rise of organized religion after 500 CE, it is not surprising that a typical European school of the early Middle Ages was rural and associated with a monastery. Thus, one might expect the early development of schools to favor theology and the teachings of the Church, which they did, rather than secular topics and study of the physical world. A broader curriculum and more open attitude in the Latin-speaking West would accompany reurbanization of Europe in the eleventh and twelfth centuries.

The eastern half of the former Roman Empire also suffered wars and economic decline after 500 CE; however, the consequences for its intellectual culture were less severe. Emergence of the Byzantine Empire provided political stability, including greater continuity in schools, where study of the classics continued in Greek. The military conquests of Alexander the Great in Asia and North Africa from 334 to 323 BCE seeded many pockets of Greek civilization, most notably Alexandria in Egypt and the Kingdom of Bactria in Central Asia, and these contributed to greater longevity of Greek culture in the East. Translation of Greek works into Arabic was accomplished largely during the eighth and ninth centuries. Although the Arabic and Islamic contribution has sometimes been described as primarily translation and preservation of Greek texts, undeniably innovative, new work, building on that of Greek scholars, was completed in the fields of astronomy, mathematics, and optics. Even so, conservative Islamic forces did hasten the decline of scientific practice within the Islamic world by the end of the thirteenth century.[6]

In time, the European universities that trace their origins to the Middle Ages would come to play a vital role in the natural sciences. Universities often arose from preexisting schools or informal associations of masters and students without the trappings of official charters and university status. Among the earliest universities were Bologna by 1150, Paris by about 1200, Oxford by 1220, Cambridge by 1225, and Salamanca by about 1230. Early universities faced the difficult challenge of securing necessary economic support, including tax relief, and political protection while maintaining all or most of their self-government and intellectual freedoms, including the ability to determine the curriculum and award degrees. This was accomplished, over time, by currying support and protection from those in power. Most teachers were also clergy members, but there was, for the most part, a remarkable degree of noninterference by the Church. The early universities were voluntary associations or guilds. They owned no property or buildings, and they could, and sometimes did, move if it improved their circumstances. The universities' programs in law, medicine, and theology in particular came to be seen as valuable assets requiring nurturance by local communities. The primary scientific subjects were astronomy, geometry, and arithmetic, although as translations of Aristotle's works on physics, meteorology, psychology, and natural history became available in the late twelfth and early thirteenth centuries, these often became compulsory subjects as well. Scholars and students at these medieval universities sought to apply Aristotelian logic to claims of knowledge, and in these universities Greek and Arabic science found, at last, an institutional home.

However, many of the leading authority figures of the time were theologians who were eager to press logic to serve their beliefs. As an early example, Anselm of Bec sought to prove God's existence through the application of reason, rather than faith alone. He did this not because he had personal doubts, but to make God's existence clear to nonbelievers. Although his "ontological proof" (the existence of the idea of a perfect being was its own proof), completed in 1078, convinced few if any nonbelievers, it did introduce a lasting problem for theologians and an unwelcome worry for secular scholars; if reason can prove theological claims, it can also disprove them.

The encroachment of natural philosophy on topics already addressed in scriptures was unpopular with many Church officials. Allegations of pantheistic teaching at the University of Paris, in which God was equated with the materials and forces of the universe, led to a decree from a council of bishops in 1210 (renewed in 1231 by Pope Gregory IX) forbidding reading and instruction on Aristotle's natural philosophy in Paris until the books were "examined and purged of all suspected error." No record of the purged version is known, and enforcement of the ban weakened around 1240, possibly influenced by the death of Pope Gregory IX in 1241 and a perception among the faculty that their reputation was suffering, relative to other universities without such restrictions.[7]

A second phase of the conflict between Church doctrine and natural philosophy culminated in the Condemnation of 1277. Pope John XXI had instructed the bishop of Paris, Etienne Tempier, to investigate the controversy surrounding alleged errors in the teaching of philosophers, including Aristotle, Averroes, Avicenna, al-Ghazali, and Moses Maimonides, and in March 1277, Tempier issued a condemnation of 219 propositions or articles.[8] They illustrate the inevitable conflict of that era between theologians, who claim authority on matters of revelation, and natural philosophers, who promote the explanatory powers of reason. Rather than condemning specific books or published statements, the condemned propositions cover a range of issues and attitudes associated with natural philosophy. The following Articles are examples of ideas that were condemned.

> *Article 40:* That there is no more excellent state than to devote oneself to philosophy.
> *Article 145:* That there is no question disputable by reason which a philosopher should not dispute and resolve.
> *Article 150:* That a man ought not to be satisfied to have certitude on any question on the basis of authority.
> *Article 152:* That theological discussions are based on fables.
> *Article 153:* That nothing is known better because of knowing theology.
> *Article 154:* That the only wise men of the world are philosophers.

Others of the articles object to any limitations imposed on the absolute power of God:

> *Article 35:* That without a proper agent, as a father and a man, a man could not be made by God [alone].
> *Article 9:* That there was no first man, nor will there be a last; on the contrary there always was and always will be the generation of man from man.
> *Article 21:* That nothing happens by chance, but that all things occur from necessity.

Interestingly, the last two articles, attributed to natural philosophers of the time (or at least some interpretations of them), are no longer tenable scientifically. Our species is not eternal, and chance events do have an important role in the physical world, from the positions of electrons, to chance mutations, to extinctions resulting from climate changes. This change indicates that the conflicts were as much about methods of knowing and intellectual territoriality as about the particular knowledge claimed.

Eleven days after Tempier issued the condemnations, a smaller but similar set of prohibitions were decreed by the archbishop of Canterbury, Robert Kilwardby, applicable to all England. Punishment for teaching prohibited articles was, as in Paris, excommunication and probable loss of employment, depending on the school. The efficacy of the condemnations is difficult to judge, though as late as 1341 new masters of arts at the University of Paris were required to pledge to teach "the system of Aristotle and his commentator Averroes, and of the other ancient commentators and expositors of the said Aristotle, except in those cases that are contrary to the faith." Of course, none of this stopped the advance of Aristotelian philosophy and its widening application. What did occur was the growing realization of the limits of logic, especially its ability to resolve theological issues. William of Ockham (1285–1347) was among those who pointed out their incompatibility. He frequently invoked the absolute freedom of God in proposing both philosophical and theological explanations. He is more familiar to some for his promotion of the medieval rule of economy, roughly stating that "plurality should not be assumed without necessity" or that the simplest explanation is preferable, which has come to be known as "Ockham's razor." Heroically applying the razor to human affairs, Ockham, when commanded to review several papal proclamations, concluded that the Pope had lapsed into heresy. His life afterward was consumed by politics.

Three with Vision: Frederick II, Francis Bacon, and Benoit de Maillet

Consideration of three historical figures, spanning the thirteenth to the eighteenth centuries, will help give a sense of the difficult birth pangs of free inquiry into the natural origins of life and its change over time. I could have chosen others, but the three discussed here are exemplary for their vision of the broad scope of natural sciences and their positive impact.

Frederick II was the last in a long line of German kings and was at various times in his life Holy Roman Emperor, the king of Sicily and Jerusalem, and founder of the University of Naples. He grew up in early thirteenth-century Sicily amid great cultural diversity, achieving competency in six languages, and combined in his studies the Greek classics, the writings of Byzantine and Jewish philosophers, and the exploratory treatises of the West. These varied influences were reflected in the wide circle of scholars and artists that he employed and encouraged. He had several of Aristotle's works translated from Greek or Arabic into Latin. Frederick was an ardent falconer and natural historian. He wrote a lengthy treatise on falconry, thirty years in the making, that included detailed accounts of avian reproductive behavior, feeding and hunting habits, annual migrations, habitat preferences, and physiology based on his own extensive observations in the wild and some experimentation with captive birds. This treatise, *De Arte Venandi cum Avibus* (published ca. 1250), is one of the earliest zoological works written in the critical spirit of modern science. Frederick maintained this critical attitude toward "all things in heaven and earth, both gods and men" throughout his life.[9]

Figure 1.2 Frederick II (1194–1250); Holy Roman Emperor, king of Sicily and Jerusalem, pioneering naturalist, patron of the arts, and foe of papal authority.

This critical attitude, together with his political power, ambition, and prickly personality, contributed to his being excommunicated, twice. He respected Aristotle, but disagreed with him frequently, stating that he relied too much on folklore and had too little direct experience to speak with real authority.

Frederick conducted experiments with vultures whose eyes were covered and nostrils plugged in an effort to determine whether vultures are attracted to carrion by smell or by sight. He raised by hand a set of nestlings found in the wild and demonstrated that some cuckoos lay their eggs in the nests of other birds and leave them to be cared for by their foster parents. He delighted in finding nests of "barnacle geese" to disprove the then-popular myth that they hatched from barnacles. His combination of keen observation with a rationalist attitude toward nature is apparent in a (prescient) statement on mating: "Nature in her endeavor to preserve the race by the continuous multiplication of individuals has decreed that every species of the animal kingdom . . . shall take pleasure in sexual union so that they may seek instinctively to bring about such enjoyment. Birds take such delight in this natural function that even birds of prey who at no other time seek the companionship of their kind, not only come together at the mating season but even exhibit definite signs of mutual affection."[10] Frederick would demand evidence for unsubstantiated explanations. Frederick spent a great deal of time in the field, and when traveling sought the company of local natural history experts. Following the crusade of 1238 Frederick brought back both falconers and birds from Arabia to learn about their techniques of hunting.

Frederick II had a lengthy dispute with the papacy based in part on the military threat he posed to the papal states in Italy and more generally on his liberal, anticlerical policies. Upon his death in December 1250, he was reviled by some churchmen, including Nicholas of Cardio, who wrote that God, seeing the desperate danger in which the storm-tossed "bark of Peter" stood, snatched away "the tyrant and son of Satan" who "died horribly, deposed and excommunicated, suffering excruciatingly from dysentery, gnashing his teeth, frothing at the mouth and screaming." Among the people, Frederick was deeply mourned throughout Europe as the *stupor mundi,* the "wonder of the world." His cultural impact was such that many could not believe he had died. False Frederick sightings were common, and stories circulated that he had gone into hiding (perhaps within Sicily's Mt. Etna) and would eventually return to reform the Church and reestablish a *pax Romana.*

Born 311 years after Frederick II's death, Francis Bacon brought the same critical attitude to bear on natural philosophy and scientific inquiry. Throughout his stormy career as a British lawyer, member of Parliament, poet, social critic, Lord Chancellor of England, and historian, Bacon was an outspoken advocate for the bright future of unfettered scientific investigation and its potential benefits to society. The *Novum Organum* (1620), or "new method," served as a rebuke to the scholastics of his own and earlier times, and the book's subtitle, "Aphorisms Concerning the Interpretation of Nature and the Kingdom of Man" leaves no doubt about his ambition. One of his aphorisms in particular (no. 49) indicates his low opinion of much of what passed for knowledge.

> The human understanding resembles not a dry light, but admits a tincture of the will and passions, which generate their own system accordingly; for man always believes more readily that which he prefers. He, therefore, rejects difficulties for want of patience in investigation; sobriety, because it limits his hope; the depths of nature, from superstition; the light of experiment, from arrogance and pride, lest his mind should appear to be occupied with common and varying objects; paradoxes, from a fear of the opinion of the vulgar; in short, his feelings imbue and corrupt his understanding in innumerable and sometimes imperceptible ways.

He indicts Aristotle, saying,

> Some men become attached to particular sciences . . . they wrest and corrupt them by their preconceived fancies, of which Aristotle affords us a signal instance, who made his natural philosophy completely subservient to his logic, and thus rendered it little more than useless and disputatious.[11]

Bacon was not a great practitioner of science, and he seemed to be unaware of current controversies, including the work of Galileo, his contemporary. Nevertheless, he was a great visionary and popularizer of the scientific enterprise at a time of cultural turmoil. Bacon added the promise of science to earlier utopian visions that sought change through social legislation, religious reforms, or the spreading of existing

Figure 1.3 Francis Bacon (1561–1626); British essayist, historian, and visionary for the scientific enterprise. Courtesy of the National Portrait Gallery.

knowledge. In his book *New Atlantis* (1624), Bacon says, "The end of our foundation is the knowledge of causes, and secret motions of things; and the enlarging of the bounds of human empire, to the effecting of all things possible."[12] He is explicit about the possible benefits of science: "We do publish such new profitable inventions as we think good. And we do also declare natural divinations of diseases, plagues, swarms of hurtful creatures, scarcity . . . ; and we give counsel thereupon, what the people shall do for the prevention and remedy of them." Bacon insisted that philosophical dogma should be re-

Figure 1.4 Frontispiece from the *History of the Royal Society,* published in 1667, showing a bust of Charles II, the Society's first patron. Viscount Brouncker, the Society's first president, appears on the left of the bust, and Francis Bacon on the right.

placed by inductive reasoning and experimental methods. Despite shortcomings, the force of his arguments and personality had considerable impact on both his peers and intellectual descendants. He was claimed as an inspirational figure by the founders of the Royal Society of London, begun as a scientific society in about 1650. Echoing Bacon, the Royal Society's motto is *"Nullius in verba,"* or "take no theory on trust."

Benoit de Maillet was a French diplomat with a keen interest in natural history, who served in Egypt and traveled extensively throughout the Mediterranean region in the late seventeenth and early eighteenth centuries. De Maillet published broadly as well on political, economic, and scientific topics. In a remarkable book, written sometime between 1692 and 1718 but published posthumously in 1748, de Maillet outlines a theory of transformation over time of the earth and its inhabitants. As indicated by the title, *Telliamed: Or, Discourses Between an Indian Philosopher and a French Missionary, on the Diminution of the Sea, the Formation of the Earth, the Origin of Men and Animals, And other Curious Subjects, Relating to Natural History and Philosophy,* he presents his unorthodox ideas through the mouth of an Indian philosopher named Telliamed (his own name in reverse) in a manner similar to Galileo's use of fictional characters in his *Dialogues.* Based on detailed geological observations made during his travels, he argues that the biblical chronology for creation could not be correct. Presaging the work of later eighteenth- and nineteenth-century geologists, de Maillet, via Telliamed, seeks to explain the history of the earth based on an extension of observable geological phenomena and the time required for change. He describes an oceanographic station designed to measure the rate of diminution of the sea and argues that all land masses consist of former marine sediments. He describes all plants and animals, including humans, as originating in the seas and becoming terrestrial through a process of adaptive transformation, following emergence of the continents.

De Maillet is rarely discussed as a significant precursor of Darwin, and I do not imply that his ideas fit a modern view of evolution. (The term "evolution" itself, little used by Darwin, did not become closely associated with the idea of descent

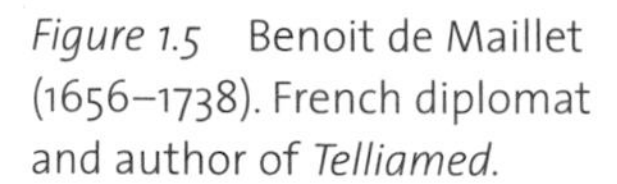

Figure 1.5 Benoit de Maillet (1656–1738). French diplomat and author of *Telliamed.*

with modification until the end of the nineteenth century.) I do suggest, however, that de Maillet, despite his many bizarre impossibilities and unsupported suppositions, struggled to provide explanations that were consistent with existing knowledge. He found clear evidence in the terrestrial sedimentary rocks and their fossils that the lands he traversed had been submerged a very long time ago, and he sought to explain the significance of his ideas for the origin and history of animals, including humans. In considering historical figures, it is important to assess their work in the context of the background knowledge of their time and the particular questions they faced. Did de Maillet suggest that different species were related to each other by common descent with modification so they might colonize newly exposed land? Arguably, yes. Consider this quote from *Telliamed:* "The transformation of a silkworm or of a caterpillar into a butterfly would be a thousand times more difficult to believe than that of fish into birds if this metamorphosis did not occur daily before our own eyes . . . The seeds of these same fish, transported into marshes, may have also originated the first transmigration of the species, from the marine to the continental environment. If 100 million individuals perished in contracting the habit, it is only necessary that two should succeed in order to start the species."[13] His discussion of mechanisms are laughable, but only in hindsight.

The opening pages of *Telliamed* find the Indian traveler responding to the missionary's question about his religion and the reasons for his travels. This passage, written over 280 years ago, provides insight into de Maillet's concern over the perceived threat within his ideas, and a timeless plea and statement about the benefits of keeping scientific pursuits free from the strictures of religion.

> Sir, I have always declined speaking to you of my religion because it can be of no use to you, and, since all men are naturally prejudiced in favor of the religion in which they are born, it offends them to have the articles of it contradicted. For this reason, and on the advice of my late father, I have all my life avoided entering into the matter in order not to generate disputes in which everyone thinks it a point of honor and conscience to support his own opinion, and which always end in nothing but mutual animosities . . . I would not even have spoken to you about my ideas on the composition of this globe, the study of which is the reason of my travels, if I had not recognized in you a mind capable of overcoming the prejudices of birth and education, and

> above all, of not being shocked by the things I intend to tell you. Perhaps they will at first appear to you contrary to the content of your sacred books, yet I hope in the end to convince you that they are not really so. Philosophers (permit me to put myself among them, however unworthy of the name) rarely find these happy dispositions . . . not even in the countries of liberty where it has been often dangerous for some of them to have dared speak against popular opinions. Besides, continued our philosopher, you have traveled a great deal . . . and you seem to think that the secrets of nature are not unworthy of your curiosity. You have learned to doubt, and any man who can do so has a great advantage over the other who believes implicitly . . . According to one of your authors, the knowledge acquired with a critical mind and by comparing the truthfulness of opposed opinions is certainly the most reliable because it is based on our reason and not on the prejudices of birth.

Telliamed was edited by de Maillet's publisher, who sought to dilute the refutations of religious dogma; however, this did not forestall either the scathing attacks by threatened theologians and clergymen or the careful consideration of later scientists, including Buffon and Cuvier.

Focus on Variation

As Enlightenment philosophers in the eighteenth century resisted any limits on the purview of human reason, especially those imposed by the Christian church, willingness grew to consider natural rather than supernatural causes for the origin of life. Evidence was accumulating that the earth was much older than the Bible dictated, and numerous fossil forms came to light that indicated extinction or change over time for species. Such evidence further fueled support for a theory of natural origins.

The geological evidence consisted of an abundance of sedimentary rock on dry land, clearly the result of accumulating sediments at the base of former oceans, lakes, or streams, and the observation that current processes of geological change would require long expanses of time to yield the observed distribution of these terrestrial sediments. Careful observation convinced nearly all naturalists that the fossils found in this rock were indeed the remains of living organisms. Yet some devout naturalists, unwilling to accept the idea that the Creator had allowed some of his works to go extinct or suffer imperfection, explained them away as purely mineral structures that had formed within the rocks.[14] Some of the fossils found in mountainous regions in sedimentary rocks were aquatic organisms. To adequately account for these, the earth would have had to be much older than anyone had thought. During the eighteenth century, the idea of change in species over long time periods became increasingly compelling to naturalists.

Carolus Linnaeus, the eighteenth-century Swedish naturalist and founder of the classification system bearing his name, is well known as a believer in the "fixity" of species and their creation by God. Less well known is his change in attitude as he gained more experience. Though originally he argued against this idea, he came to accept the possibility that new species might arise through the action of different environments. His change of heart was based primarily on his field observations of extensive morphological variation within species, particularly for plants. Linnaeus was an avid gardener and was quite successful in the production of hybrids. Not surprisingly, in his *Disquisitio de Sexu Plantarum* published in 1756, Linnaeus presented the possibility that hybridization between two species might give rise to a third. Linnaeus suggested that God may have created a single species as the foundation for each genus and then allowed the multiplication of species to proceed via hybridization. By the time of the publication of the twelfth edition of his major work, *Systema Naturae,* in 1766, he had dropped his earlier assertion that no new species arise.

Georges de Buffon, in *De la Degénération des Animaux,* also published in 1766, supported the idea that similar species shared a common ancestor and diverged in appearance and habits over time. Further, Buffon saw that geographic distribution provided a clue regarding the history of organismal or species diversification, with populations becoming divided through migration to different regions of the world, eventually yielding different forms. His notion of species was imprecise, but precise species definition remains difficult. Buffon divides the history of the earth into seven distinct periods of change, with man appearing last, a fact that indicates the evolutionary tendencies of his views. In his essay "Époques de la nature," he names the first period "when Earth and the plants were formed," the fifth "when elephants and other tropical animals inhabited the North," the sixth "when the continents were separated from one another," and the seventh "when man appeared." Buffon was one of the first to dedicate himself to investigation of the earth's history and to consider the development and lives of organisms within this context. Buffon read and commented on the works of de Maillet and also served as mentor to the young J. B. Lamarck.

Jean Baptiste Pierre Antoine de Monet, Chevalier de Lamarck was born in northern France in 1744. His parents sent him to a Jesuit school and hoped he might become a priest. Instead, during the latter part of a long and embattled career as a natural scientist, he provided the first detailed theory of organic evolution, in a series of works published between 1800 and 1822. Lamarck's scientific interests were as broad as Buffon's before him, though it was zoological work, particularly in classification and systematics of invertebrates, where he was most successful. In *Zoological Philosophy* he wrote, "The aim of a general arrangement of animals is not only to possess a convenient list for consulting, but it is more particularly to have order in that list which represents as nearly as possible the actual order followed by nature in the production of animals: an order conspicuously indicated by the affinities which she has set between them." He continues, "Thus to obtain a knowledge of the true causes of that great diversity of shapes and habits found in the various known animals, we must reflect that the infinitely diversified but slowly changing environment in which the animals of each race have successively been placed has involved each of them in new needs and corresponding alterations in their habits." Further, "There is one strong reason that prevents us from recognizing the successive changes by which known animals have been diversified and been brought to the condition in which we observe them; it is this, that we can never witness these changes. Since we see only the finished work and never see it in course of execution, we are naturally prone to believe that things have always been as we see them rather than that they have gradually developed."[15]

Lamarck provides an explicit diagram "showing the origin of the various animals." Of course, many of his hypotheses are wrong. For example, he supported two separate origins for animals: one for "worms" and their descendants and one for "Infusorians, Polyps and Radiarians" and their descendants. He placed monotremes, which include the duck-billed platypus, as more closely related to birds than mammals. He got many things right, however, including placing cetaceans (whales and dolphins) correctly as mammals. Lamarck reasoned that loss of the limbs and the pelvis in whales "is the result of an abortion due to long disuse of them." Lamarck relied largely on the consequences of use and disuse, including the heritability of newly

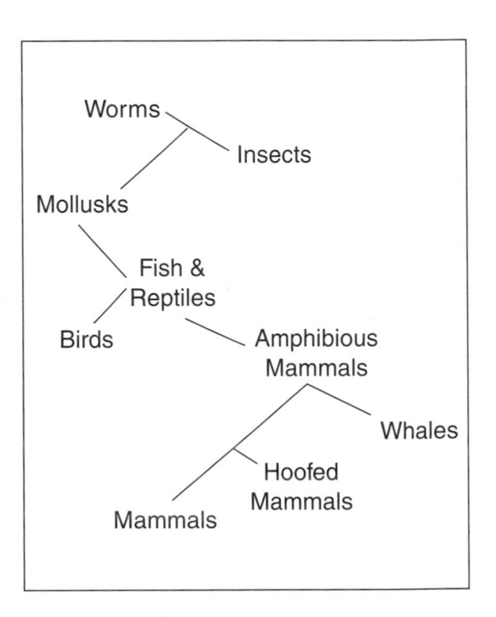

Figure 1.6 Evolutionary tree from J. B. Lamarck's *Zoological Philosophy* (1809).

acquired characteristics, as the mechanism causing change in organisms. He states, "By the influence of environment on habit, and thereafter by that of habit on the state of the parts and even on organization . . . any animal may undergo modifications, possibly very great, and capable of accounting for the actual condition in which all animals are found."[16] The notion of inheritance of acquired characters has become known as "Lamarckism" and is discredited (though in rare instances it may apply). But this idea was commonly held by naturalists during the late eighteenth and nineteenth centuries.

Lamarck did not hesitate to include humans within his scheme of shared ancestry among animals in general and primates in particular. He equivocates only at the end of a short passage on the development and diversification of human languages, saying, "Such are the reflections which might be aroused, if man were distinguished from animals only by his organization, and if his origin were not different from theirs."

This sentence provides some cover from accusations of heresy, though few doubted that Lamarck included humans in his scenario for life's natural origins and diversification.

Lamarck's ideas about species transformations convinced few, if any at all. He was prone to assertion rather than provision of detailed evidence, and the mechanism of inheritance of acquired characters could not be demonstrated or sustained. Lamarck's contemporaries had much greater appreciation for his insights into the classification of invertebrates than his views on the mutability of species. Some clergy and devout philosophers felt threatened by his views; however, because Lamarck had little support from his peers, they could dismiss his work as hopelessly materialistic. Today Lamarck is seen increasingly as a seminal figure, providing the "first major evolutionist synthesis in modern biology"[17] and a springboard for discussions of organic evolution in Charles Lyell's *The Principles of Geology, Vol. 2* (1832) and Robert Chambers's *Vestiges of the Natural History of Creation* (1844). Lamarck was famously abused in the eulogy at his own funeral delivered by his influential rival, Georges Cuvier. Cuvier, a gifted anatomist and the founder of paleontology, scorned "the fanciful conceptions" and indulgent "imagination" that he found in Lamarck's works, though Cuvier's constrained vision limited him to the fated view of the immutability of species.

Erasmus Darwin, a physician and the grandfather of Charles Darwin, believed God had designed living forms to be self-improving through time, with new capabilities and organs developed by means of the inheritance of acquired characters. In *Zoonomia,* published in 1794, he wrote, "Would it be too bold to imagine, that in the great length of time since the earth began to exist, perhaps millions of ages before the commencement of the history of mankind . . . that all warm-blooded animals have arisen from one living filament . . . with the power of acquiring new parts, attended with new propensities, directed by irritations, sensations, volitions, and associations; and thus possessing the faculty of continuing to improve by its own inherent activity, and of delivering down those improvements by generation to its posterity?" Erasmus Darwin's ideas regarding evolution were only briefly presented, but they indicate an awareness and circulation of the idea of the changeability and development of species over time, apart from understanding of mechanisms.

The turning point in modern biological and evolutionary understanding finally arrived on 24 November 1859 with publication of *Origin of Species* by Charles Darwin. His book has two themes: (1) common descent with modification for all forms of life, and (2) modification of groups of organisms via natural selection.[18] He marshaled disparate evidence on geological processes, age of the earth, variation in traits within and among geographic populations of organisms, factors affecting individual's reproductive success, artificial selection on domestic species, and extinction and habits of fossil forms. He showed how descent with modification could explain them all. Darwin went on to publish six editions of the highly successful *Origin of Species* and nine other books further developing his ideas.

Cultural historians may credit Darwin with the idea of evolution, but this would be inaccurate, as the idea, in its early stages, had been disseminated in the West at least since the writings of de Maillet in the mid-1700s. Darwin did provide the mechanism and the arguments that finally convinced the scientific and educated lay communities of the fact of common descent for all organisms. Darwin had little but public silence and private disdain for Lamarck. In correspondence with Charles Lyell in 1859, Darwin stated, "Plato, Buffon, my grandfather before Lamarck, and others, propounded the obvious view that if species were not created separately they must have descended from other species, and I can see nothing else in common between the *Origin* and Lamarck. I believe this

[Lamarck's] way of putting the case is very injurious to its acceptance, as it implies necessary progression, and closely connects Wallace's and my views with what I consider, after two deliberate readings, as a wretched book, and one from which . . . I gained nothing."[19] For Darwin, common ancestry for species was obvious, and his key contribution lay in explaining the mechanism of natural selection. For this, the second theme of his book, acceptance came more slowly. Although natural selection is now widely accepted as a primary mechanism for evolution, lively debate continues as to its relative influence and timing for particular traits. Somewhat later, in the preface of the sixth edition of *Origin of Species,* Darwin acknowledged the relevance of many earlier or contemporary authors and credited Lamarck for "arousing attention to the probability of all change in the organic, as well as in the inorganic world, being the result of law, and not of miraculous interposition."

Darwin was the original and most comprehensive explicator of natural selection, though even here he was not entirely alone. Two decades of careful study and voluminous correspondence on the topic were brought to a forced culmination (in publication of *Origin of Species*) when Darwin received a short manuscript from Alfred Russel Wallace in 1858. Based on his own extensive field work on insects in the Malay archipelago, Wallace briefly outlined a theory of organic evolution by means of natural selection similar to Darwin's. Wallace is properly regarded as a co-discoverer of natural selection and a pioneering evolutionist. However, Wallace felt natural selection ("survival of the fittest") could not explain features seemingly unique to humans, including "calculation of numbers, ideas of symmetry, of justice, of abstract reasoning, of the infinite, [and] of a future state," and he invoked the action of "an Overruling Intelligence . . . to aid[ing] in, the indefinite advancement of our mental and moral nature." For this restriction on natural causation, as well as for his defense of séances and other spiritualistic happenings, Wallace has puzzled even those who admire his insights.[20] It was Darwin who blazed the trail and first made an extensive case for organic evolution in a consistent and compelling manner. Within Darwin's lifetime the fact of evolution became widely accepted by the scientific community.

Our understanding of biological systems has advanced dramatically since Darwin's time, and organic evolution, rather than withering under the onslaught of new data, has become the indispensable and unifying concept used by all biologists. A small sampling of material discoveries since Darwin's time, enabling and further explaining evolutionary processes, includes examples from many scientific fields.

Discovery of the natural radioactive decay of uranium from one of its forms (isotopes) to another in 1896 by Henri Becquerel, a French physicist, heralded the application of physics to measurement of the age of the earth's materials. Experiments over the next twenty years showed the time scale available for evolution to be billions of years, not just thousands. On another front, Gregor Mendel's work, published but little appreciated during his own lifetime, was rediscovered around 1900, demonstrating some traits (color and shape of plant parts) to be heritable as discrete units, persistent over time, rather than continually blended across generations. DNA was identified as the material of inheritance in 1944, and its physical structure explained in 1953. Thus, one or a few nucleotide changes could be seen as the basis for some phenotypic variation among individuals. The code by which DNA specifies amino acids was elucidated in the 1940s and 1950s, and over the next half-century was found to be nearly universal across life forms. This is in keeping with common ancestry for all life forms, and the small number of differences found are shared by close relatives, indicating evolvability of the code itself. Conceptual and empirical development of the field of population genetics throughout the twentieth century demonstrated how rare changes can become fixed in populations by both natural selection and chance. The longstanding puzzle of altruistic behaviors can be addressed in terms of kin selection theory, in which the proportion of genes shared among interacting individuals is taken into account.

Discovery of sea-floor spreading during the 1960s, vindicating the earlier continental drift theory of Alfred Wegener, enabled biogeographers to better explain current distributions of both surviving and extinct groups of animals and plants. An abundance of intermediate fossil forms have been discovered—two of the best-known examples being *Archaeopteryx,* a small dinosaur with feathers, teeth, and tail vertebrae, and *Basilosaurus,* a whale with legs.

Discovery of the common developmental role of *hox* genes in defining animal body segments, and uncovering the elaborate history of *hox* gene duplication and divergence underlying body plan differences, provides an observable, physical basis for understanding complex change in body design. Horizontal transfer of genetic material between species, mediated by viruses or by symbiotic relationships, has been discovered, and this provides a qualitatively different mechanism for evolutionary change. Many occurrences of horizontal gene transfer, both ancient and recent, are only now being uncovered.

With the aid of large data sets, powerful computers, and improvements in analytical methods, the study of phylogenetic relationships has succeeded in providing explicit demonstration of the hierarchical relationships of groups of organisms in nature. Analyses of disparate data sets including both DNA and morphology are broadly congruent. The molecular sequence for functionally important shared genes now allows systematists to include representatives of all organismal groups, from bacteria to humans, in simultaneous analyses of their shared evolutionary history.[21] This puts hypotheses of the pattern of descent with modification on firm empirical grounds, with objective criteria and methods that provide reproducible results.

Studies in experimental evolution have taken organisms with short generation times, such as bacteria or viruses, into controlled laboratory conditions and documented their evolutionary change. This includes imposing speciation events by separating groups of individuals (moving them into different flasks) and forcing them to evolve separately. Experiments such as this provide dramatic demonstration of the ability of populations to evolve differences appropriate to their particular environments and capacities quickly, in a period of months or weeks rather than geologic eras. One of the lessons from these evolutionary experiments is that methods for inferring the phylogenetic relationships among taxa do indeed work. Analyses of the DNA sequences collected from different individuals over time identified the same patterns of phylogenetic relationships that had been experimentally imposed.

This chapter has traced the history of three initially unpopular discoveries. These are: the fact that the earth orbits the sun, the fact that many diseases arise naturally from microbial life forms, and the fact of common ancestry for all organisms. All three of these discoveries presented essential challenges to orthodox thinking and traditional institutions at the time of their formulation. Heliocentrism removed the earth from the center of the universe and contradicted the Bible. The germ theory of disease origins removed a category of direct punishment or reward from the diminishing arsenal of divine power, although it must have seemed, at first, similarly intangible and even more unjust. Common descent for all organisms knocked humans from their elevated position as the intentional and favored creation of a divine being, wholly distinct and superior in origins, capabilities, and destiny from other forms of life.

All three also required extrapolation from observable phenomena to causes operating at vastly different scales of distance or time, and observable only in part or indirectly. For heliocentrism, distant planetary movements spanning long periods of time had to be envisioned based on fragmentary data. In the case of germ theory, people had to infer, at first, the presence of small infectious agents from their physiological effects and geographic distributions alone, and, in the case of evolution, people had to use present-day observations of geology, biological variation, and differential survival to piece together events of the earth's history and life's history from the far distant past.

Natural philosophers and scientists pursued these ideas not to discredit particular institutions or the existing cultural order (though that may have been partial inducement for some), but to inquire, because inquiry comes so naturally and can lead to one of the deepest satisfactions that humans experience—an improved understanding of the world. However, this does not mean that scientific ideas, once they attain a degree of acceptance, are complete and protected from further testing. If anyone can provide compelling evidence to modify or overturn generally accepted ideas, that would ensure that person acclaim (at least, eventually) for having succeeded where others have failed.

Acceptance of new ideas tends to follow demonstration of their ability to explain observations and the presence of a suitable cultural climate (both intellectual and social). It is not enough for an idea to be conceived, it must be pursued, tested, and applied, if possible. Acceptance of the three unpopular discoveries discussed here were subject to these requirements. Which theory met with the most resistance, either from the scientific community or the general public? Evolution seems a likely candidate. This may well be because its implications are most immediate to our understanding of the human condition.

How did we originate? Why do we exist? Why do we act as we do? What has been and what is our role in the world and in society? What are we capable of accomplishing? At some level these are all evolutionary questions. Choosing a precise date for the origination of an idea is difficult, and it is perhaps even more difficult to choose a precise date of acceptance of that idea. Nonetheless, attempting an estimate would be a valuable exercise.

The original scientific formulation for heliocentrism may be dated to about 1510, the time at which Copernicus' *De revolutionibus* was written, though it was not published until 1543. Acceptance among scientists may be placed roughly after the death of Galileo in 1642, and acceptance by the primary community resisting the idea may be estimated, in the extreme, as 1835, when Galileo's *Dialogues* was removed from the "Index of Prohibited Books," published by Catholic authorities. This gives an estimate of 325 years for acceptance by its most reluctant audience.

We can take the publication date for Girolamo Fracastoro's *On Contagion* in 1546 as initiating germ theory discussion, and 1886, following the convincing work of Robert Koch that linked specific microbes to specific diseases, as the date for both scientific and public acceptance. This yields an estimate of 340 years for germ theory from first explicit formulation to widespread acceptance. Though the gestation period was long, germ theory moved quickly to public acceptance following clear and repeated demonstration of the link between microbes and disease and the benefits this brought for human health.

If we take 1718, the approximate time of de Maillet's writing of *Telliamed*, as the initiation of evolutionary thought, and 1859, the publication date of Darwin's *Origin of Species* as the time of acceptance by the scientific community, this yields an estimate of 141 years for acceptance by scientists. Estimating a date for acceptance of evolution by religious authorities is inevitably subjective, and will vary across and even within groups. We might consider the statements of the popes as a conservative measure. In 1950, Pope Pius XII stated a willingness to consider evolution of "the human body," though he felt that possibility remained to be determined. Pope John Paul II's statement, delivered to the Pontifical Academy of Sciences in October of 1996, goes further. It can be taken as an acceptance of human evolution from other forms of life based on acceptance of a credible series of discoveries and their independent confirmation. He said, "My predecessor Pius XII had already stated that there was no opposition between evolution and the doctrine of the faith about man and his vocation. Pius XII added . . . that this opinion [meaning evolution] should not be adopted as though it were a certain, proven doctrine . . . Today, almost half a century after the publication of the [1950] encyclical, new knowledge has led to the recognition of the theory of evolution as more than a hypothesis. It is indeed remarkable that this theory has been progressively accepted by researchers, following a series of discoveries in various fields of knowledge. The convergence, neither sought nor fabricated, of the results of work that was conducted independently is in itself a significant argument in favor of the theory."[22] Debate among Catholic clergy will continue; however, a precedent for their church's acceptance of human evolution has been set. If 1996 is used as an estimate for a general acceptance (though many would disagree) and 1718 for an early formulation for descent with modification for species, we have an estimated span of 278 years.

By this reasonable though admittedly subjective accounting, acceptance for evolution has not taken any longer than for heliocentrism or germ theory. The ship of cultural change, though slow to change course on evolution, appears no slower in changing than for other discoveries of similarly high social impact. This may be a surprising and possibly welcome perspective for educators in the United States fighting a rearguard battle with certain religious fundamentalists over the teaching of evolution in public schools.

It is my premise in this book that the fact of evolution (descent with modification for all life forms), long accepted by scientists, has many applications that are deeply integrated into our lives and our societies, often in ways that we do not realize. The study of evolution and the application of the growing knowledge of evolutionary mechanisms has inevitably become a necessary endeavor for biologists. Our understanding of evolution is also increasingly relevant to human welfare, however. It is the goal of this book to outline some applications of evolutionary biology to the important topics of individual health, public health, agriculture, conservation of natural resources, resolving criminal court cases, and to a limited extent providing a context and metaphor for human cultural affairs.

2

DOMESTICATION: EVOLUTION IN HUMAN HANDS

Though it took 300 years or more for the idea of evolution to gain widespread acceptance, humans have been applying the evolutionary concept of natural selection in strategic ways for well over 10,000 years. By promoting the survival and reproduction of some individuals over others within populations of various animal and plant species, humans have influenced the evolution of size, shape, physiology, behavior, and environmental tolerance within those species. Sometimes the changes have been subtle, accumulating slowly over time. Other times they have been dramatic and fast, as chance mutations provided traits seized upon and favored by savvy humans.

Are the changes wrought by humans intentional or simply accidental? Certainly there have been many chance events, but just as certainly there have been concerted, intentional efforts to maintain particular changes serving human needs for food, clothing, shelter, work assistance, and companionship. Just as a hunter need not recall the mechanics of gravitational theory to estimate the right trajectory for an arrow, a gatherer need not conceive the mechanistic details of evolution to preferentially collect and sow the seeds of the best food plants. The hunter understands the arrow will fall, and the gatherer understands that individuals vary and like begets like. Application of these biological facts of life have, literally, been our bread and butter.

The term "artificial selection" is often used to describe the domestication process whereby humans select certain individuals for breeding and not others based on the presence of desired traits. To call the process artificial is misleading, in that all animals impose selection at some level on the species they consume, and humans are no more artificial than tigers. However, we do protect favored individuals from the pressures of natural selection experienced in the wild, and in doing so, promote some genotypes and traits that would be disadvantageous in the wild.

We trace the origin of our species, *Homo sapiens,* back at least 140,000 years ago to fossils found in Africa that are characteristically human in appearance. These early humans were resourceful hunters and gatherers of the woodlands and open habitats. The climate during the long expanse of the Pleistocene (1.6 million to 10,000 years ago), when humans arose, was highly changeable and characterized by periods of extreme glaciation and cooling interspersed with much warmer periods, like the one we are currently experiencing. As a result, early human populations periodically dispersed to higher latitudes, including Eurasia, and were left to move, die out, or cope as best they could when conditions changed. It was the descendants of these colonists who began activities (such as hand-raising young animals taken from the wild and sowing seeds of favorite plants) that led eventually to agriculture and domestication of some animal and plant species.

A warming climate at the end of the Pleistocene, concomitant with increasing human population sizes and technological advances, set the stage for the independent development and spread of domesticated species in multiple geographic centers, including the Fertile Crescent, China, Mesoamerica, Andes and Amazonia, eastern United States, Sahel, tropical West Africa, Ethiopia, and New Guinea. Different progenitor species of plants and animals were necessarily pressed into service in different regions. The Fertile Crescent, including the Tigris and Euphrates rivers, encompassed present-day Iraq in the east to the Mediterranean Sea in the west. It provided the natural habitat and was the native home for wild peas, barley, and two kinds of wheat, as well as bezoar goats, Asiatic mouflon sheep, wild boar, and aurochs, which served as progenitors for domesticated goats, sheep, pigs, and cows, respectively. China was home to wild rice, millet, soybean, and the Bactrian camel. In the New World, the Andes provided the initial breeding stock for domesticated lima beans, peanuts, potatoes, manioc, and squash, as well as llamas and guinea pigs.

No doubt the factors favoring successful development of domesticated species varied across these regions, although ultimately both human need and opportunity were involved. The disparate locations mentioned share the key feature of being the native home of wild species especially well suited to domestication, due to a combination of chance and adaptation to local conditions. The earliest sites for domestication within these different regions also tended to be populated by relatively affluent, sedentary societies who used a variety of wild food species in their diets. They lived near marshes, lakes, or rivers and tended well-watered, fertile soils. Our understanding of the origins of agriculture and the domestication of plant and animal species has grown in recent years with the complementary efforts of archeologists, anthropologists, geneticists, and evolutionary biologists.[1]

In this chapter I illustrate some of the history of evolutionary change that humans have wrought through domestication. I contend that the intentional and readily observed changes demonstrate what can be considered the first and most important application to nonhumans of our growing understanding of the biological facts of life, including evolution.

Domestic Animals and Plants

Dogs

Humans created dogs from wolves. Wolves have been our partners, but, as with all domestications, we have driven the process. Wolves and dogs can and still do interbreed, and since 1993 have been considered a single species by the International Commission on Zoological Nomenclature. The innate sociability of both wolves and humans, including their shared use and acceptance of dominance hierarchies, made theirs a natural association. During the Pleistocene humans and wolves would have known each other well, where their ranges overlapped, as competitors in hunting prey, as mutual threats, and as occasional mutual beneficiaries. Humans likely provided some food for wolves following human encampments, and wolves provided companionship and protection by warning humans of the approach of strangers. Other canid species are less well suited to a human association. Their habits are more solitary, or they use fewer mutually recognizable cues for communication. For example, although African hunting dogs *(Lycaon pictus)* are social and hunt in packs, their dominance hierarchies are less pronounced, and social status is often signaled by the passing of regurgitated food rather than visual behavioral cues such as posture, which is used and readily understood by wolves and humans. Natural predisposition to successful communication between wolves and humans has even been found to be selectable and improved in dogs. Tests of the cognitive ability of dogs and wolves to use such human cues as tapping to indicate the location of hidden food and pointing found dogs to be more successful.[2]

The process of domestication can be seen as a natural series of interactions ranging from mutual observation by wary antagonists and competitors, to mutual benefit in finding or sharing food, to rearing and taming of individual wolves by humans, to differential survival and breeding of wolves based on heritable variation in their physical and behavioral traits. Overly aggressive wolf pups would be killed by humans keeping them, whereas pups and adults adapting readily and deemed beneficial to human groups would be favored. Artificial selection by humans was likely preceded by some natural selection among wolves based on variation in their innate abilities to tolerate and adjust to humans. Those best able to adapt to life on the fringes of human society, perhaps as camp followers, were likely among the first lineages to be intentionally bred by humans.

A well-known series of experiments beginning in 1959 by Dmitri Belyaev and colleagues used silver foxes *(Vulpes fulvus)* to show how quickly human-imposed or artificial selection can work.[3] Starting with wild-behaving foxes on fur farms, they selected for and bred individuals showing reduced levels of aggression and fear of humans. After only eighteen generations, tame behavior had become the norm within the experimental population, and unexpectedly, new traits, mirroring those seen in many dogs, began to appear, including patchy coat coloration, curly tails, and drooping ears, as well as reduced levels of corticosteroids and increased levels of serotonin. The latter hormones play roles in inhibiting stress responses and aggressive behavior, respectively. All these traits, seen in the artificially selected silver fox adults, resemble traits seen in wild juveniles to some degree. This supports the idea that the evolution of domestic dogs has occurred, in part, through the re-

Figure 2.1 Evolution of dogs.

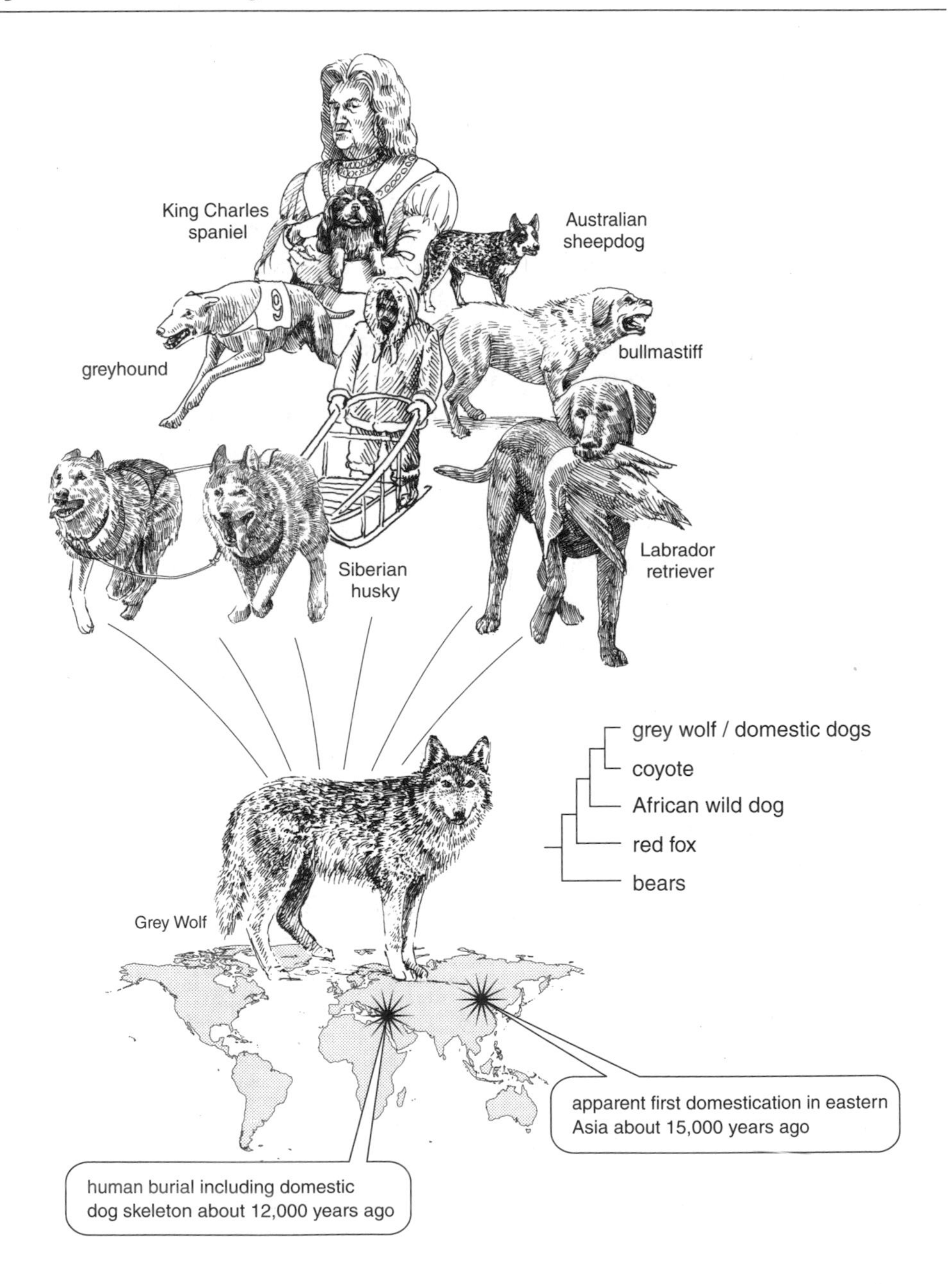

Figure 2.2 Evolution of horses.

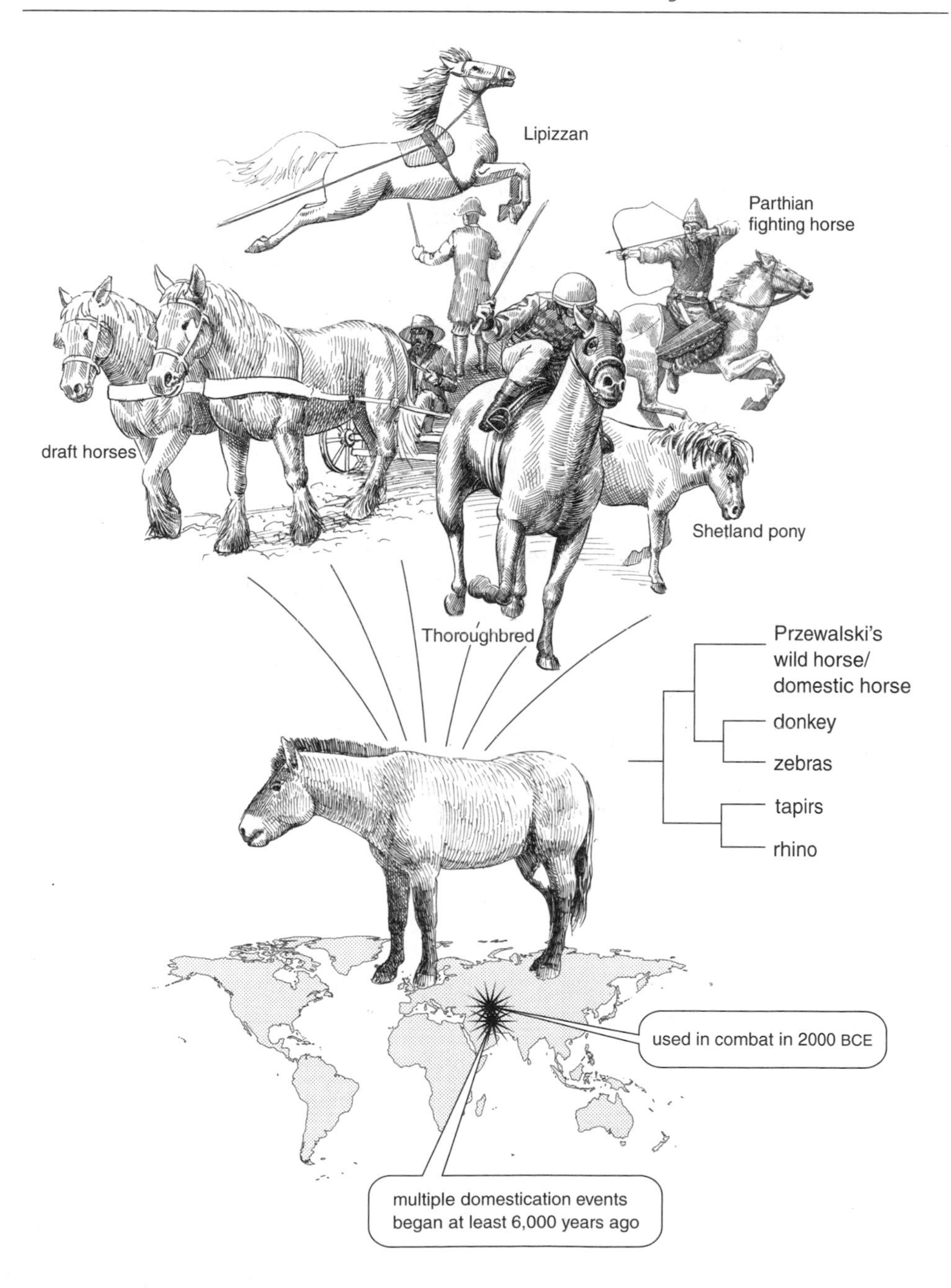

tention of juvenile features in adults, a change in developmental timing known as "neoteny." This would also explain the retention in most adult dogs of juvenile-wolf skull proportions and possibly certain juvenile behaviors, including the barking, fawning, and face-licking well known to pet owners.

Current molecular evidence suggests that dogs were first domesticated in east Asia, and that all breeds are descendants from this first set. This is based on the observation that mitochondrial DNA diversity is greatest among domestic dogs generally native to east Asia as measured against a worldwide sample of 654 dogs. The estimated date of origin is about 15,000 years ago. This is determined using a fossil-based calibration of 1 million years ago for the divergence of wolves and coyotes, and an estimate of the rate of substitution per million years calculated by comparing mitochondrial DNAs among extant wolves and coyotes. Ancient dog remains exhumed from Alaskan permafrost and from Latin America, which predate the arrival of European settlers in the New World, were found to be genetically very similar to European and Asian dogs, suggesting they were also descended from wolves domesticated initially in eastern Asia that traveled to the New World across the Bering land bridge with some of the first human settlers of the Americas over 12,000 years ago.[4]

The earliest archaeological evidence of dogs includes dog-like jaws from 14,000 years ago discovered in central Europe and skeletal remains of a puppy buried 12,000 years ago, in today's Israel, in a stone-covered tomb under an elderly woman's arm. Dog remains can be distinguished from wolf remains by the relatively small jaw size, the tight spacing of premolar teeth, and their close association with humans.

Evidence indicates a trend toward increasing variation in dog morphology over time. For example, dog remains from 2000 BCE in England are similar to each other in size with a shoulder height of about 50 centimeters (cm), whereas dogs are more variable in size during the Iron Age about 700 years later. Dog remains from Roman Britain during this period suggest a range in shoulder height from 23–72 cm.[5] Based on stone carvings and pictures in ancient Egypt and western Asia from at least 2000 BCE, there was great variation in dog sizes and shapes much earlier than noted in England, with some of the pictured dogs bearing strong resemblance to modern mastiffs and greyhounds.

The phenomenal range of variation in size, shape, temperament, and behavior within modern dogs has arisen within the past 15,000 years; however, a great deal of that range is much more recent, being only a few hundred or even a few decades old.

The historical record of the classification of dog breeds gives some indication, though certainly incomplete, of human efforts in the development of domestic dog diversity. In the late 1500s John Caius, a physician and anatomist, classified dogs into three groups: curs, hunters, and companions, with curs including shepherds, mastiffs, and bandogges. About two centuries later Carolus Linnaeus published the names of thirty-five different dog breeds. Today, as many as 700 different breeds of dog are claimed or recognized, though "breed" is an arbitrary distinction. Determining the phylogenetic relationships among breeds is difficult or impossible owing to frequent mixing among them and the small amounts of genetic differentiation and time involved.

And what have humans made of the wolf, over millennia of selective breeding, training, trading, and cohabitation? Consider the seven categories of dog breeds currently recognized by the American Kennel Club (AKC) as a rough guide. Hounds have been intensively selected for their acute sense of smell. This includes the bloodhound, so named not because of a stalking blood lust, but because of extensive efforts in pure breeding of these "blooded hounds." They were favored by the Monks of St. Hubert monastery in Belgium, and their pure-blooded ancestors date back at least to the eleventh century. Sporting dogs, like the Brittany with its sensitive nose and ready obedience, have been bred to work closely with people in hunting birds. With training, they can be directed to search fields for quarry, freeze, point to and flush live birds from cover, and then retrieve them once brought down. Working dogs are a diverse set, bred to serve as guards like the Doberman pinscher, pull sleds like the Alaskan malamute, or assist in water rescue like the Newfoundland. Terriers have been bred to hunt rats, badgers, and otters, excavating burrows and ripping through brush as necessary. So-called toy dogs have been bred largely for companionship, of the valuable nonjudgmental sort. One example, the Pomeranian, has been intentionally selected over the past 150 years for smaller size and sociability from much larger sled dogs of Iceland. Herding dogs, selected for the ability to assist in herding livestock, include breeds such as the Australian cattle dog and border collie. The Australian cattle dog is the result of extensive cross-breeding over the past sixty years of the dingo (a breed that some believe originated in Australia as long as 4,000 years ago), the blue smooth highland collie, the Dalmatian, and possibly the bull terrier. The nonsporting dogs include the bulldog, used during the 1600s in pit-fights against chained bulls or bears and bred for tenacity and ruthlessness. Since the 1830s when such fights were banned in England, bulldogs have been transformed via selective breeding into more gentle and personable show dogs. With intentional and continual selection for traits by humans, wolves have been adapted for diverse abilities and a place of prominence in human culture.[6]

Horses

Domestic horses are descended from the wild horses that ranged widely across the Eurasian steppes during much of the

Pleistocene, about 2 million to 10,000 years ago. Wild horses that were hunted for food are depicted in cave paintings in Europe that are as much as 30,000 years old. Horse domestication began at least 6,000 years ago, based on the presence of bones and of molars indicating bit wear, possibly associated with riding, at archaeological sites in Ukraine and Kazakhstan of the Eurasian steppes. The earliest written and artistic evidence of horse domestication activity dates from over 4,000 years ago.

Molecular phylogenetic analyses of diverse modern horse breeds and Pleistocene archaeological materials for wild horses indicate that there were multiple, independent domestication efforts beginning with different wild horse populations.[7] This is based on the presence of greater genetic diversity among some modern domestic horse breeds than between some of those modern horse breeds and horse remains from late Pleistocene archeological sites. The first efforts likely involved hand-rearing by humans of infants born to wild parents. Foals may have been taken after mothers were killed by hunters. As with wolves, the natural social behavior of horses, including their acceptance of dominance hierarchies, facilitates their interaction with humans.

The earliest efforts in domesticating horses likely centered on herding and eating them. Piles of chopped horse bones found along with other kitchen debris at multiple archaeological sites support this notion. Stunning evidence for the use of horses in combat is found in grave sites dated to about 2000 BCE on the Eurasian steppes near the border between Russia and Kazakhstan, where human warriors are buried together with teams of horses and two-wheeled chariots (as well as bronze knives, sickles, and arrowheads). Charging horses dragging war carts with spear-wielding warriors must have been an intimidating sight upon first encounter by Bronze Age peoples. The Hittites used horse-drawn chariots to conquer Mesopotamia and Egypt around 1800 BCE, and horses remained a vital component in human wars for over 3,700 years.

One of the last cavalry attacks may have been as unexpected by the victims as the first attacks by Hittites on their southern neighbors. In 1945, a Polish cavalry brigade fighting alongside the Soviet Union in World War II stormed a line of German tanks and anti-tank gunners. Emerging from the smoke of burning Soviet tanks, the horsemen took the German gunners by surprise and were able to move past the line, dismount, and attack from the rear.[8] Some of the Polish cavalry's horses were said to be descendants of the horses of the ancient Scythians, skilled horsemen and warriors whose domain ran from central Europe across the steppes of southwest Asia from the eighth to the fourth centuries BCE.

During the fifth century BCE the Persian Empire was administered and maintained with the crucial aid of a mounted messenger fleet. It was about these horsemen that Greek historian Herodotus wrote, "Neither snow, rain, heat, nor darkness stays [them] from the swift completion of their appointed rounds." It was with the vital aid of horses that the Romans built 50,000 miles of highways running from Britain in the west to Syria in the east. The exploits of the Spanish conquistadors in the New World during the 1500s were greatly facilitated by their importation of horses. Stock farms established in Panama in 1514 provided the twenty-seven or so horses that Francisco Pizarro first took to Peru in 1531. In 1539, Hernando De Soto brought an army of about 620 men and 220 horses to Florida. In 1540, Francisco Vázquez de Coronado brought about 1,500 horses, along with cattle and mules, on his expedition seeking the rumored seven cities of gold.

The horse-drawn Conestoga wagons were instrumental in settlement of the western United States by European settlers. Small horses (ponies) were used to haul coal from mines starting in the early 1600s. In some of the large mines in Europe, the pit ponies were bred, born, and worked entirely below ground without ever seeing daylight. Shetland ponies were bred by Shetland islanders, off the north coast of Scotland, to haul peat for fuel and seaweed to fertilize the fields. They were imported to Britain in the 1800s to haul coal in the mines, and were imported again to the eastern United States when coal was discovered in West Virginia and Kentucky. From antiquity up to the invention of the internal combustion engine just a century ago, horses and horse-power were generally the most efficient means of travel, hauling goods, tilling soil, and waging war.

Intensive selection and breeding of horses during the past several hundred years or so has resulted in a great diversity of distinctive breeds, including over 200 that are currently recognized. Work horses include the Belgian, descended from the Flemish horses favored by knights as a battle horse in the thirteenth century, and now bred as a powerful draft horse. Racing horses include the Thoroughbred, which is the result of intensive selective breeding for competitiveness, grit, and speed since the 1790s.

Cattle

Cattle have been the single most important domesticated animal for humans as a source of meat, milk and milk products, and manure for fertilizer. They have proved invaluable as a work animal assisting in plowing farm fields, hauling goods, and transporting humans. The earliest archaeological evidence indicating domestication of cattle dates to about 6200 BCE from Turkey. There are now about 800 recognized breeds, all descended from wild aurochs *(Bos primigenius)*. Male aurochs stood six feet high at the shoulders and had large, forward-curving horns. Aurochs were the subjects of cave paintings in Europe over 30,000 years ago. They survived in central and western Europe until the Middle Ages, and the last were killed in Poland around 1627.

Phylogenetic analyses of mitochondrial DNA for diverse cattle breeds indicate two separate origins for domestic cattle: one for Indian cattle, and another, separately, for European cattle, both from wild aurochs.[9] Each of these two lineages has many closely related breeds, but the European and Indian groups as a whole are most distantly related to each other, with an estimated divergence time of just less than one million years ago, indicating their separate origins in different geographic areas prior to domestication. European cattle were first domesticated in western Eurasia and Mesopotamia, whereas Indian cattle, with distinctive humps not found in European cattle, were domesticated in eastern Eurasia, ranging from Iran to India.

Archaeological finds from 5000 to 2500 BCE indicate that humans selected for a dramatic reduction in size of domestic cattle, and some little more than a meter in height at the shoulder are known from sites dating to the first millennium BCE. Smaller size would have made cattle easier to handle and house, and may underlie the observed trend. More recently, livestock breeders have developed highly specialized breeds such as Holsteins, originally bred in the Netherlands and selected for milk production. Holstein cows yield an average of forty-five pounds of milk per day and remain productive for about six years. Herefords, originally bred in Herefordshire, England, for meat, have bulls weighing, on average, 1,900 pounds. These breeds, intensively selected and managed for high yields of milk and meat, are generally maintained on high-quality diets in carefully controlled environments. Other cattle breeds were selected for multiple uses and survival in more extreme conditions. The N'Dama breed, originating in equatorial west Africa, yield about five pounds of milk per day, and males average about 650 pounds. N'Damas, however, have an extremely important advantage over nonnative breeds with higher milk or meat yields. They have evolved resistance to trypanosomiasis, a debilitating and often fatal disease of cattle caused by small protozoan parasites (trypanosomes) spread by tsetse flies. In this instance, artificial and natural selection both contribute to a locally well-adapted, and economically important, domestic breed.

Whether or not cattle were bred and used for their milk has, in turn, influenced the evolution of human physiology. With the exception of certain human populations, milk consumption among mammals stops after weaning. This is accompanied by a decline in production of lactase, the enzyme that digests milk sugar. Human populations without a tradition of milking and milk consumption beyond early childhood, as in large sections of central Africa and eastern Asia, become naturally lactose intolerant, due to low or no production of lactase, as they age. Human populations with a long history of milking and milk consumption in adulthood have evolved regulatory gene changes that extend lactase production, and lactose tolerance, into adulthood.

Chickens

Domestic chickens have been developed by humans from one or more lineages of red junglefowl *(Gallus gallus),* a species native to tropical rainforests of southeast Asia. Red junglefowl live in social groups of about thirty individuals, and the brightly colored males compete in performing ritualized displays to attract females. Human hunters and gatherers apparently learned early on that removing one or more eggs from the nests of junglefowl induces females to lay more, thereby providing more to gather. Archaeological evidence suggests that domestication dates to about 5200 BCE in China and 3200 BCE in India. Chickens appear on Assyrian coins from the

Figure 2.3 Evolution of cattle.

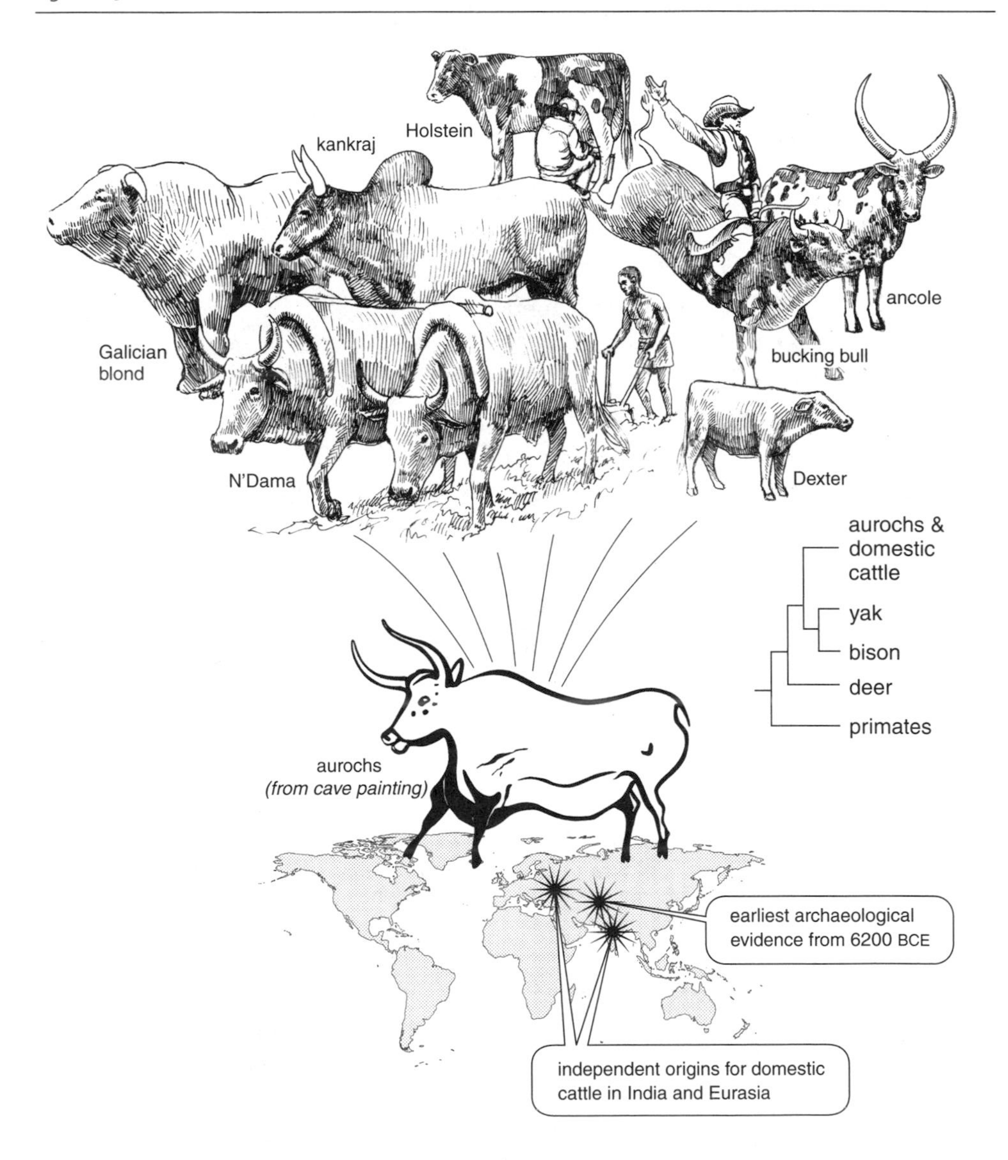

Figure 2.4 Evolution of chickens.

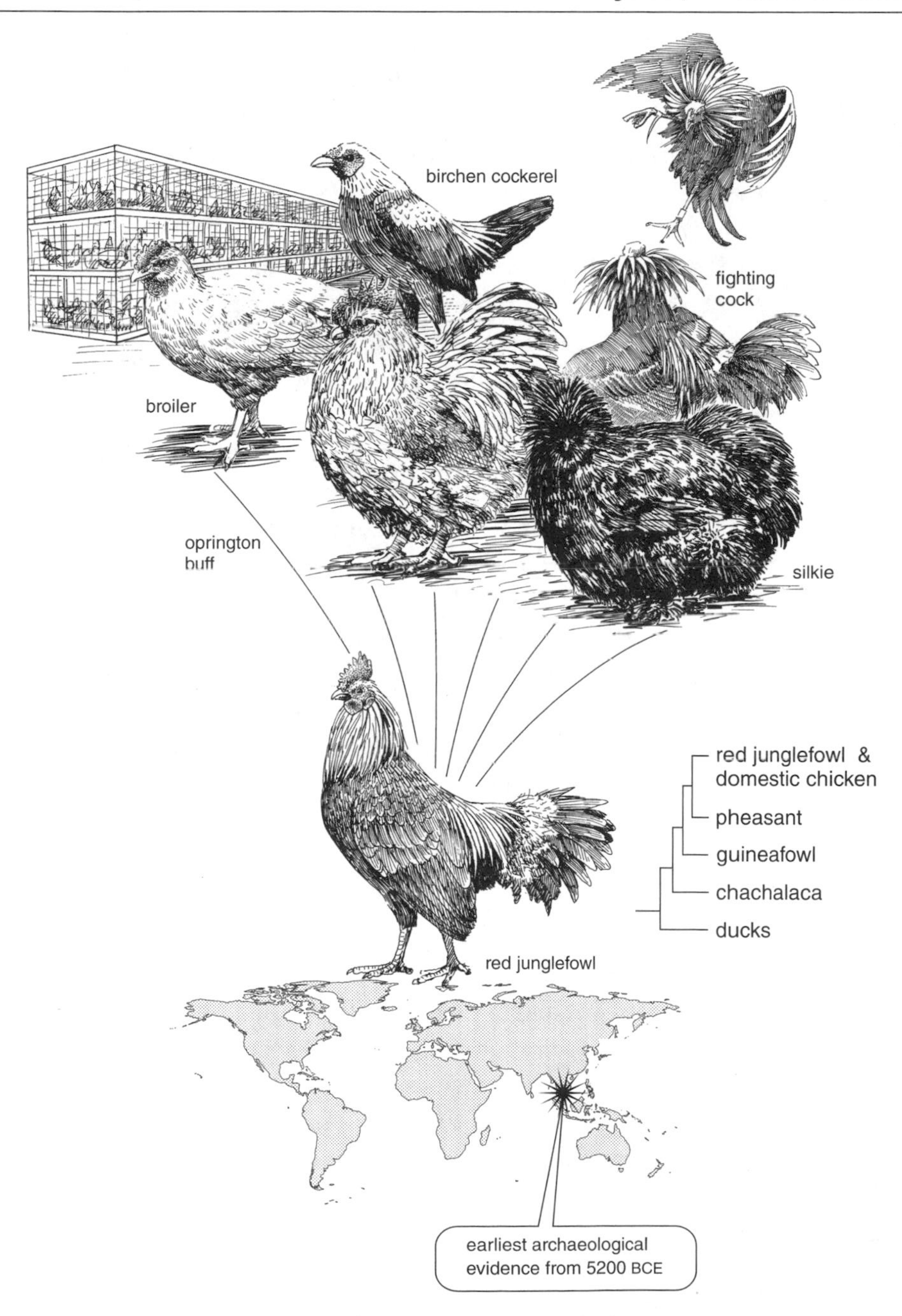

eighth century BCE, and by the fourth century BCE, the Greeks were carefully breeding chickens for meat and egg laying as well as for their fighting abilities. Columella, writing in Latin in the first century CE, noted the prized breeds of fighting cocks in Italy—the Rhodian, Chalcidian, and Median. Offering this evidence of intensive breeding, Columella recommended that no more than fifteen eggs be placed under a single hen at one time lest she have trouble successfully incubating them all simultaneously. The Roman naturalist Pliny wrote that to hasten fattening, meat chickens were placed in cages so small that their heads and tails protruded at each end and they could do nothing but eat. Varro, another author of the same period, wrote that some farmers fattened hens with wheat bread soaked in wine. He described how the skin and feathers of chickens infested with fleas or mites were treated with ash produced in making lye soap, and detailed the need and operation of nets to keep hawks from entering chicken pens.[10]

During the first century BCE, and likely earlier, Egyptians built incubators of clay bricks and warmed them by carefully tending fires to approximate the body temperature of a brooding female (ca. 105° F). Reportedly, thousands of eggs were incubated simultaneously and young chicks kept warm, for directly after hatching they could not move about readily on their own.[11]

Chickens provided key components for many ancient folk medicines. Hippocrates prescribed egg whites mixed with water as a laxative. Pliny recommended a cocktail of chicken eggs, raisin wine, and starch to restore virility in men, and (of course) chicken soup for stopping dysentery. Such recommendations, together with recipes for cooking chickens "stuffed with cooked peas, brains, and Lucian sausages, seasoned with freshly ground pepper, lovage, oregano and ginger, and basted in olive oil and wine" helped the Romans popularize chickens through much of Roman Europe. In about 1605, Ulisse Aldrovandi, a professor at the University of Bologna, published fifty pages on medicinal uses of chickens of the time.

The modern poultry industry began in the nineteenth century and grew dramatically after World War II with the advent of mass production facilities containing rows of pens and stacked cages of birds. The world's chicken population has recently been estimated at over 17 billion, and Americans consume more than 100 pounds of chicken per person per year. About sixty different breeds of domestic chicken are recognized. White leghorns, originating in Leghorn, Italy, are the most numerous breed in North America and are most often selected for egg production. Today's broiler chicken attains a body weight of about 4.5 pounds in just forty-two days, and layers yield 200 to 300 eggs per year. This easily beats the wild red junglefowl, who weighed about half as much and laid five to ten eggs annually.

Other Animals

There are other domestic animals with histories similar in outline but different in their particulars. Domestic hogs have been developed from wild boar of Eurasia and northern Africa; domestic carp from wild carp separately in both China and Europe; tilapia from wild cichlid fishes as long as 4,000 years ago in Egypt; and domestic turkey from the wild turkey of the Americas. Domestic bees have been developed from wild progenitors for both honey and beeswax on multiple occasions, with the earliest known evidence of hives from Egyptian paintings dating to 2400 BCE. Artificial selection for honey yield, reduced aggression, early spring development, and effectiveness in pollinating specific crops has had remarkable effects, especially following development of artificial insemination methods for queens that would naturally mate with a succession of males (drones) twenty kilometers or more away from her home colony.[12]

Wheat

Grain from grasses intensively selected by humans has been the primary crop and source of calories for most civilizations around the world. Different wild grasses provided source stock in different geographic regions for independent domestication events. Wheat and barley were developed as cereal crops in parts of Africa, Asia, the Middle East, and Europe. Rice was developed in south and southeast Asia, maize (corn) was domesticated in the Americas, and sorghum and millet originated in Africa, south of the Sahara.[13] Humans were gathering and eating wild Emmer wheat as early as 19,000 years ago, based on finds of wheat seeds at an archaeological site by the Sea of Galilee.

Wild wheat and many other wild grasses are primarily self-pollinating. This is the result of pollen being shed inside the florets before they open. Frequent self-pollination can provide genetic isolation for populations and enable protection of the desired, artificially selected traits from being lost by widespread outbreeding with wild progenitors in the same region. Thus, self-pollination, arising by natural selection or chance, predisposed grasses with that capability to success as an early human cultivar. Occasional natural cross-pollination in these same grasses allows for the inflow of new genetic variants and their recombination with existing genotypes. Subsequent rounds of self-pollination can lead to the rapid fixation of the new variants, particularly if those variants are selected for by growers.

Domestication of wheat and barley allowed hunter-gatherers over time to settle in one place and to become farmers. The grains of these grasses are high in nutrients and protein and store well for long periods if kept dry. The grasses can complete their entire life cycle in a matter of months. These grass traits facilitated dramatic increases in human population sizes and the development of agricultural villages. Grain storage allowed people some reprieve from the incessant demands of food-getting, such that other skills could be developed and other activities pursued, such as metallurgy, building, writing, and trading.

Five wheat species, all in the genus *Triticum* and all self-pollinating, are generally recognized. Einkorn wheat *(T. monococcum)* has one or two grains per spikelet and was domesticated about 9,000 years ago following selection for larger seeds that stayed longer on the plant. Though widely cultivated in the past, it is a relic today. Emmer wheat *(T. turgidum)* initially had seeds with a hull; however, variant forms in which the hulls are readily removed by threshing have been favored and further developed by artificial selection. Timopheev's wheat *(T. timopheevi)* has both wild and domesticated forms, though the cultivar, developed in Georgia by the Black Sea, is no longer in widespread use. *T. uratu* is a wild wheat. Bread wheat *(T. aestivum)* is the most widely cultivated species. Based on molecular data, bread wheat arose as a result of hybridization of *T. turgidum* with a wild grass, *Aegilops squarrosa.* This hybrid was selected for certain desirable traits by humans as early as 4700 BCE, such as having hulls and grains that are easily separable from each other and from the stems in threshing. The processes of hybridization among disparate forms and of increase in the number of chromosomes via duplication (sometimes of entire genomes) is a common method of speciation in plants, and the increase in genetic material allows for the evolution of novel traits and protein functions. Humans have been able to capitalize on this tendency by identifying, selecting, and maintaining the desirable traits of bread wheat that arose initially as natural mutations in hybrids.

Figure 2.5 Evolution of wheat.

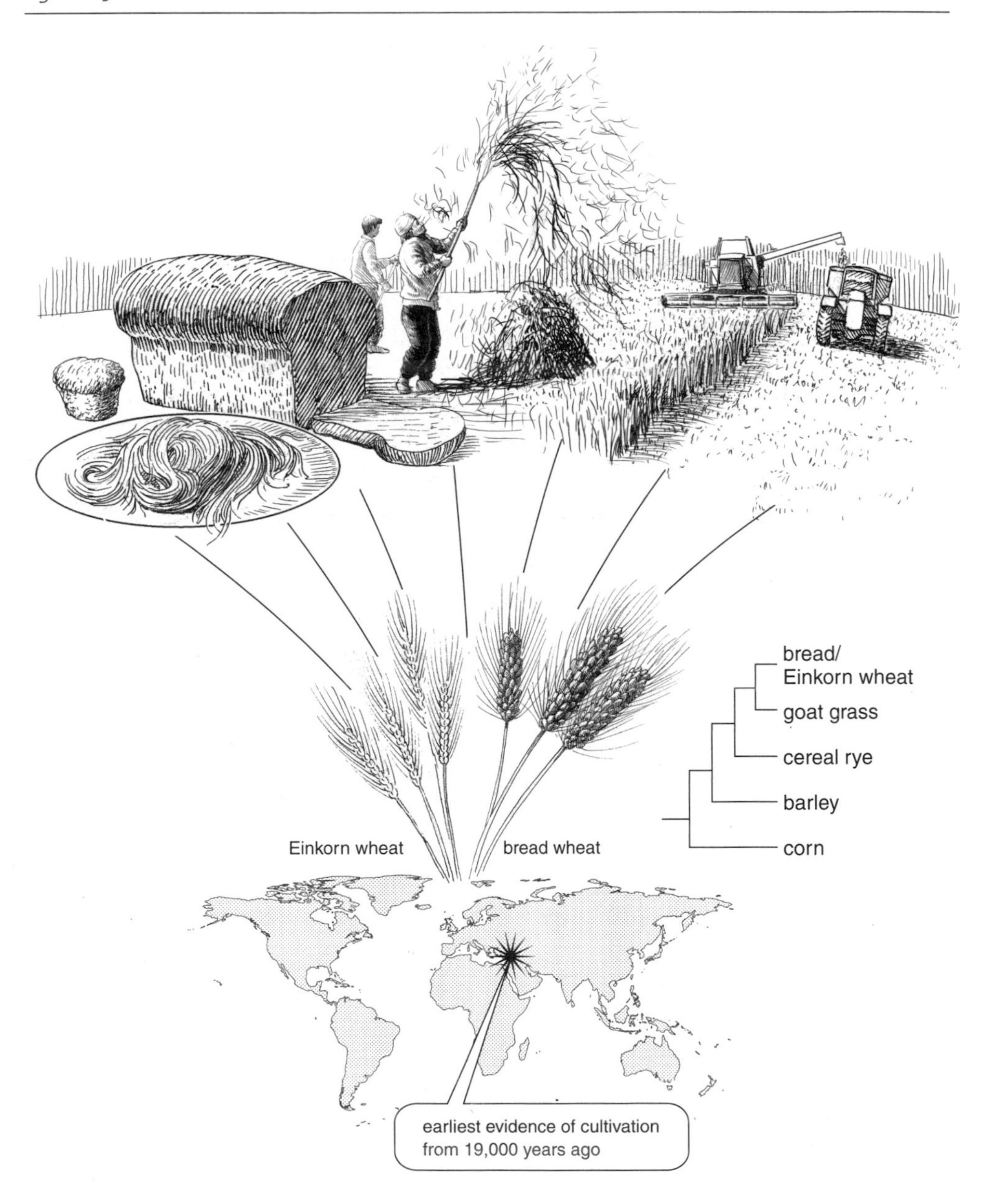

Figure 2.6 Evolution of cabbage.

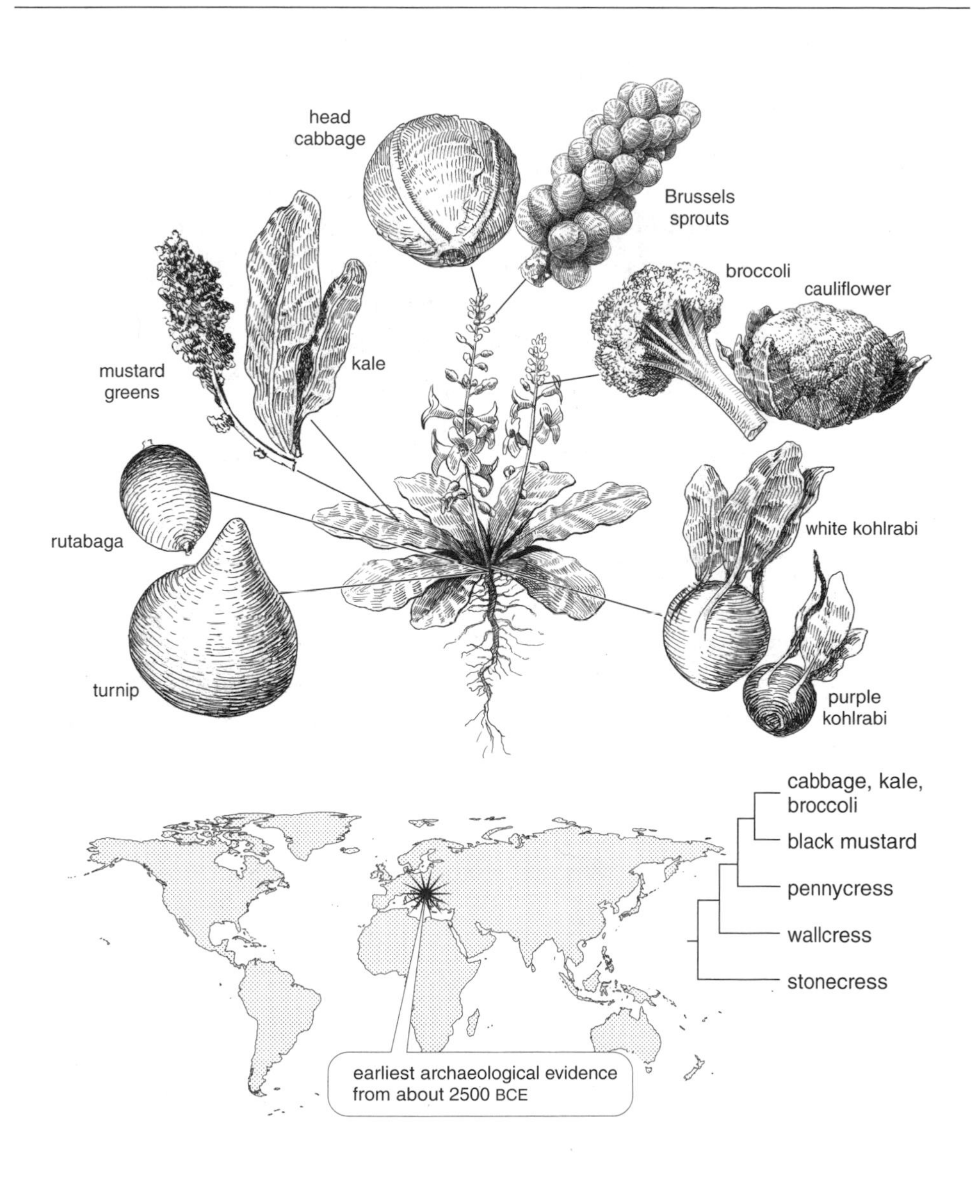

About 600 million tons of wheat grain is grown annually around the world, and over 100 million tons of that grain is sold for export. Competition among growers for market share is intense, spurring governments to invest in transport systems, to engage in continual trade negotiations, and to conduct research and development of new wheat varieties for niche markets. For example, among *T. aestivum* varieties, hard red winter wheat is favored in bread production, hard amber durum wheat is often preferred for pasta, and hard red spring wheat is generally used to produce a blend used in pastries.

Domestication of cereal grasses, and wheat in particular, has been a linchpin in the evolution of human civilization, and it came about through the pragmatic application of an evolutionary truth. Differential reproductive success of individuals, as controlled by humans, can fix desirable traits in species over time. Work on genetically modified wheats with traits such as enhanced tolerance for herbicides has already begun, and these efforts may be seen as the continuation of the artificial selection process with new technology that was started by our ancestors nearly 20,000 years ago as they collected the fruits of wild wheats.

Cabbage

The powers of artificial selection are impressively displayed by the malleable cabbage, with different parts of ancestral cabbages having been selectively favored and developed in different descendant lineages. Native to maritime habitats in the Mediterranean basin and the Atlantic coast of Europe, the wild cabbage *(Brassica oleracea)* has been cultivated since about 2500 BCE in southeastern Europe. Theophrastus wrote in the third century BCE about three kinds of cabbage in the gardens of the ancient Greeks. Based on both morphological and molecular traits, wild *Brassica oleracea* remains the closest wild relative of the following vegetables derived by artificial selection on alternative body parts.

The familiar "head" cabbage is a greatly enlarged terminal bud of *B. oleracea,* found in both red and green forms, and has been selected for large size and cold-hardiness. Literary references suggested head cabbage, commonly used in making sauerkraut, was cultivated in what is now Germany from about 1000 CE. Brussels sprouts are cultivated from the enlarged lateral buds. A single Brussels sprout plant can yield a hundred or more separate miniature cabbage heads. Broccoli and cauliflower are derived from the stems and immature flower clusters, and kohlrabi and kale (whose varieties from the southeastern United States are known as "collard greens") are derived from artificial selection accentuating the leaves.

Another *Brassica* species, *B. rapa,* very closely related to *B. oleracea,* is widely cultivated for its swollen hypocotyl (stem of the embryo), which we know as the turnip. Additional *B. rapa* varieties, many apparently first domesticated in Asia, include the Chinese white cabbage, bok choy, pak choi, bok celery, celery mustard, white mustard cabbage, plain Chinese cabbage, celery cabbage, napa cabbage, hakusai, wong bok, and Peking cabbage. Additional *B. rapa* varieties are selected for their seed-oil. The rutabaga is another artificially selected *Brassica* species *(B. napus).* Rutabaga is believed to be a turnip-cabbage hybrid originally cultivated in Europe in the 1600s. Selection for success in northern climates has facilitated its cultivation in Britain and Scandinavia. Who would have thought cabbages to be so emblematic of human achievement in the realm of evolution?

Coffee

Caffeine is a plant alkaloid that serves as a natural defense against herbivores, by way of its toxic effect on the nervous

Figure 2.7 Evolution of coffee.

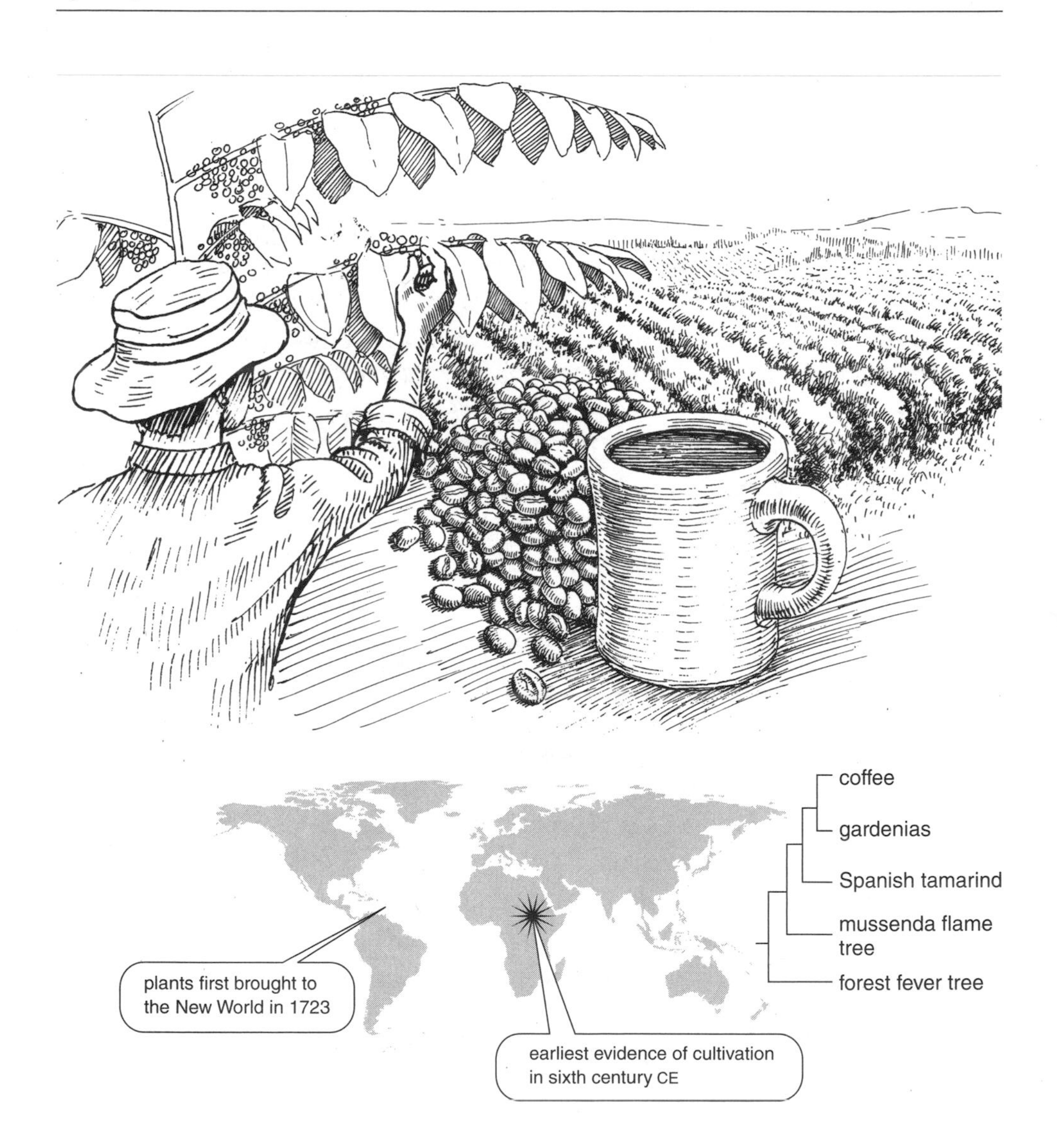

systems of would-be predators. Indicating its utility to plants, caffeine is produced by more than sixty species, including those providing coffee, tea, cacao (chocolate), kola nuts, yoco bark, guayusa leaves, yaupon leaves, maté, and guaraná nuts. Coffee beans alone provide about 54 percent of the world's caffeine kick, followed by tea and soft drinks.[14]

Coffee production relies mainly on two species, *Coffea arabica* and *C. canephora. C. arabica* is considered to produce better coffee and constitutes about 70 percent of all coffee produced. Wild *C. arabica* has its center of genetic diversity, and apparent center of origin, in the highlands of southwestern Ethiopia and adjacent Sudan. Similar to bread wheat, *C. arabica* is a polyploid (tetraploid in this case, having double the genetic material and chromosomes compared to diploids) and is self-pollinating, whereas other *Coffea* species are diploid and generally self-incompatible. Genetic analyses of *C. arabica* phylogeny and origins indicate that it arose as a result of hybridization between *C. eugenioides* and *C. canephora,* or types closely related to them.[15]

The berries from *C. arabica* may have been used initially as food, though the stimulative effect of its caffeine could not have been missed. *C. arabica* was first cultivated in southern Ethiopia during the sixth century CE. At about the same time, Arab traders carried coffee to Arabia, and coffee, as a drink originally brewed from green unroasted beans, became popular. Cultivation of coffee plants was practiced in Yemen by 1000 CE, and coffee use, including roasting and grinding the beans before boiling them in water, spread throughout the Islamic world, including northern Africa, and throughout the Middle East by 1500 CE. By the late 1600s and early 1700s coffee use had spread throughout Europe, the Caribbean Islands, and colonized portions of the Americas.

Coffea breeders have selected for plants based on their productivity, timing of fruit maturation, bean qualities, and resistance to pathogens. It is difficult to satisfy all these criteria simultaneously, however. A fungus *(Hemileia vastatrix)* known as coffee leaf rust first appeared in *C. arabica* plantations operated by British growers in Ceylon (Sri Lanka) in 1869 and spread, with devastating effect, to the plantations of southeast Asia. The fungus first infects leaves, turning them black, and eventually denudes and kills the plant. This fungus epidemic crushed the British coffee enterprise and triggered a search for resistant coffee plants. *Coffea canephora,* native to central, equatorial Africa, was both resistant to the fungus and prolific under cultivation, although the taste was considered less desirable. This species, promoted with the name *robusta,* also grew in wetter, warmer habitats at lower elevations than *C. arabica.* By 1900 *C. canephora* was cultivated on many plantations in central and western Africa and southeast Asia.

Decaffeinated coffee was first made in 1906 by treating steam-heated beans with a solvent (benzol) to extract the caffeine. Since that time, several other processes have been invented to decaffeinate beans, but although decaffeination reduces the adverse side effects of caffeine that some people experience, such as headaches and high blood pressure, it also removes flavor compounds. Recently, a gene known as "caffeine synthase" has been identified with a crucial role in caffeine synthesis, and insertion of this gene, in the wrong orientation, into *Coffea* gamete genomes may yield a plant producing less caffeine. A more traditional, nontransgenic approach to development of decaffeinated coffee may prove more successful, however. A Brazilian research group assayed 3,000 *Coffea arabica* trees and identified three naturally occurring decaffeinated mutants.[16] Using these three plants and a traditional artificial selection approach, cross-breeding the best-tasting with the least-caffeinated plants, it may be possible to develop a better-tasting decaffeinated cup. Drinking coffee for the taste and not the kick may seem strange to some, but the economic incentives are real.

Other Plants

Many other plants have been artificially selected and customized by humans. For example, over 4,000 years of human cultivation and selection of cotton (genus *Gossypium*), beginning in the Old World, has produced an increase in the length (from 20 mm to 4 cm or more) and elasticity of cotton seed fibers, enhancing their suitability for clothing. Artificial selection on annual legumes cultivated for their seeds (pulses), such as lentils *(Lens culinaris)* and peas *(Pisum sativus),* was concomitant with selection on cereal grasses and resulted in a series of traits desirable in cultivars, but potentially disadvantageous in the wild. This includes evolutionary changes in pulses toward longer retention of seeds in the pod, thinner seed coats, loss of seed dormancy, larger seeds, and stiffer stems, facilitating their cultivation as free-standing crops.

Domestication of fruit trees, such as apple, pear, cherry, and plum, is more complicated, as it requires augmenting the plants' natural sexual mode of reproduction to include a vegetative mode of propagation, accomplished artificially by the rooting of twigs or by grafting. This was required to maintain new, desired traits of size and taste, because fruit tree species in the wild maintain high levels of genetic variation. Consequently, trees raised from seeds vary widely regardless of the characteristics of the parents. Grafting of apple trees, an early form of cloning, was practiced by the Greeks as early as the third century BCE.[17] There are now over 2,000 apple varieties selectively bred for differences in taste and optimal growth conditions, with a quadrupling in size from wild apples collected by humans and found at archaeological sites dating to the third and fourth millennia BCE. The inventiveness and pragmatism of these early horticulturalists in taking a section of a stem with leaf buds from a desirable fruit tree and inserting it into the stock of another lacking the desired traits is impressive.

Unintended Effects

Unfit Design

Both artificial and natural selection proceed by differential reproductive success. However, by artificially boosting the odds of success for domesticated individuals and populations, we have accelerated and skewed the processes of selection such that our preferences supersede those favored in the wild. Many health problems for domestic animals result from intensive artificial selection for traits desired by the breeders. Some of these problems stem from compromising the once efficient design of wild progenitors, honed by millions of years of natural selection. Evolutionary theory predicts, and common sense dictates, that individuals and populations artificially protected from the rigors of survival will develop health problems. Without the filter of natural selection, mutation and recombination of genotypes yielding poorly adapted (for the wild) forms can multiply.

In dogs, some traits selected for as part of the official breed standard, often for purely aesthetic reasons, are decidedly unhealthy. Breeds with artificially selected short muzzles, such as pugs and boxers, often suffer from breathing and sinus problems directly related to their unnaturally short snouts. The reduction in hind limb and pelvis size of German shepherds preferred by AKC judges compared to the more natural proportions seen in non-AKC German shepherds lead to problems of hip dysplasia and rheumatoid arthritis. The disproportionately shortened legs and long back of basset hounds, thought to be descended from dwarfed bloodhounds, render them prone to spinal disc problems and elbow dysplasia. The long drooping ears of many dog breeds, including basset hounds, predispose them to continual infection with bacteria, such as *Otitis externa,* diagnosable by their stench. This is not to pick on basset hounds, which have many positive attributes, including a great nose, intelligence, and a gentle disposition. To cite an example in cattle, Holsteins, Herefords, and Angus are all known to suffer from accelerating aging and disease as a result of intensive selection for early, high reproductive output.

Decreasing Genetic Diversity and Biodiversity

The quest for optimal traits has been accomplished by intensive inbreeding, in which closely related individuals are continually mated to each other. Sometimes siblings are mated or parents are paired with offspring. Because the genetic variation held by entire populations and species cannot be found in any single individual, the net result of inbreeding is a loss of genetic diversity in the breeding lineages. As genetic diversity and alternative alleles (variant forms of the same gene) are lost from breeding populations, homogeneity increases, including incidence of homozygous recessive alleles with negative effects. Combining increased incidence of less-fit alleles with artificial protection from the rigors of life in the wild is a recipe for trouble.

Returning to dogs, there are many maladaptive traits associated with various breeds that have increased in recent times as a result of mutations spreading rapidly in small, inbred populations. Bernese mountain dogs suffer from a higher rate of cancer than many other breeds of the same size and similar life span. Selective breeding of the Shetland sheepdog (sheltie) to enhance the length and color of the coat in this former work dog has been accompanied by a higher incidence of pancreas problems. Doberman pinschers are susceptible to narcolepsy, a sudden descent into sleep, occasionally accompanied by paralysis. Australian shepherds show greater levels of inherited epilepsy than found in most other breeds. A series of canine diseases, recently found to be autosomal recessive traits (expressed when both parents carry the recessive mutation), can be increased by inbreeding. For example, half of all Bedlington terriers in the United States and Britain are carriers of canine copper toxicosis, an autosomal recessive disorder causing retention of copper from food that results in severe liver damage. Basenjis, all of whom are descended from dogs that came from Zaire in the 1930s, are prone to a potentially fatal autosomal recessive form of anemia, and progressive retinal atrophy in Irish setters is also an autosomal recessive disease that leads to blindness. Larger dog breeds tend to have shorter life spans, averaging six to ten years, compared to smaller breeds, averaging ten to fourteen years or longer. The reasons for this are not well understood; however, one hypothesis is that larger breeds tend to be more inbred than smaller breeds, with particularly large sires and their progeny contributing disproportionately to the population's gene pool. The unintended byproduct of such inbreeding is shorter life spans.

In horses, heritable diseases associated with inbreeding include dwarfism and narcolepsy in miniature horses, hyperkalemic periodic paralysis in quarter horses, severe combined immune deficiency syndrome (SCID) in Arabians (autosomal recessive), and overo lethal white syndrome in paint horses. In meat chickens, intensive artificial selection has been accompanied by obesity, erratic ovulation in hens, and reduced ability to fertilize eggs in males.

Intensive artificial selection improving the commercial value of plants also often leads to decreased genetic diversity and reduced ability to adapt to environmental change, including reduced disease resistance. Large population sizes, combined with low genetic diversity of many crops, makes them more abundant targets for parasites or pathogens, as those parasites pursue their own biological imperative to reproduce. The decimation of coffee trees from infection with leaf rust mentioned previously is one example. Another is the Irish Potato Famine of 1845 to 1849. Potatoes were infected with the water mold *Phytophthora infestans.* Because potatoes in Ireland were imported, they were more susceptible to infection. Having evolved in the New World and only brought to the Old World by the Spanish about 500 years ago, potatoes had no previous exposure to the Old World molds and hence no opportunity to develop resistance. Epidemic diseases continue among crops, as seen in the spread and increased pesticide resistance of pathogenic fungi of tobacco, corn, and other crops.

Spreading Pathogens

In our 10,000-year practice of animal husbandry and domestication, we have unwittingly initiated a series of evolutionary interactions between microbes and hosts, including ourselves. Some have had major impacts on the course of human history, and all illustrate the truism of evolutionary descent with modification. By living and working in close association with animals and by increasing their, and subsequently our own, population sizes, we have created abundant opportunities for microbes to disperse, adapt, and successfully colonize humans and other new host species.

Native Americans in the New World were both intentionally and unintentionally infected with smallpox viruses, tuberculosis bacteria (though some strains may have been present already), measles, and other pathogens in the conquest and settlement of the New World by Europeans. Most Europeans had sufficient experience with these pathogens, especially cowpox and its human varieties, to have evolved a degree of immune or genetic resistance to the pathogen entirely lacking in American natives. Given the relative ease with which smallpox viruses can spread, Native American populations suffered huge losses even before their face-to-face battles with Europeans began. Smallpox viruses can persist in the environment for months or years without human hosts, and hitchhiked readily on the blankets and other trade items first passed from Euro-

Figure 2.8 Evolution of smallpox viruses.

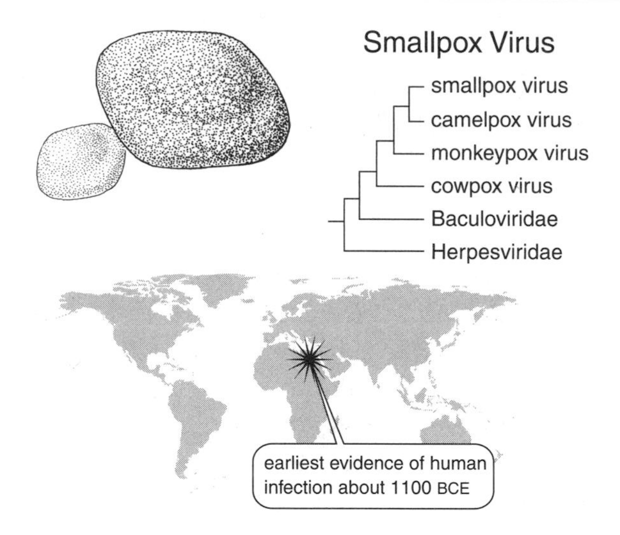

Evolution & Adaptation

- ***Evolutionary history:*** most closely related to camelpox and monkeypox viruses and placed with them in the double-stranded DNA virus family Poxviridae
- ***Adapted to:*** thrive in dense human populations
- ***Dispersal:*** humans are the only natural host; transmission by direct contact among humans or by contact with contaminated objects
- ***Recent evolution:*** last naturally occurring smallpox infection was in 1977 in Somalia; pathogen appears extinct (outside of lab samples) following global vaccination program
- ***Minimizing pathogen success in humans:*** vaccination

peans to Native Americans and then passed among the Native Americans themselves.

Perhaps 50 million or more Native Americans died during the colonial era as a result of microbes whose origins in humans can be traced to domestication of cattle. Measles in hu-

Figure 2.9 Evolution of bacteria causing tuberculosis.

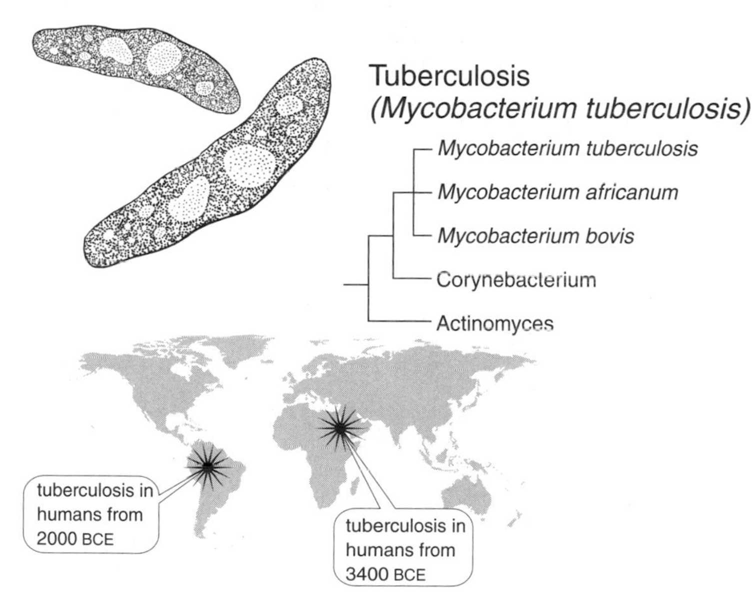

Evolution & Adaptation

- ***Evolutionary history:*** closely related to bacteria infecting lungs of cattle and other animals
- ***Adapted to:*** potential long latency periods with reactivation following weakening of host's immune system
- ***Dispersal:*** transmitted by coughing and breathing of contaminated air
- ***Recent evolution:*** increased resistance to antibiotics and increased virulence in some strains
- ***Minimizing pathogen success in humans:*** isolate infected humans; reduce selection favoring resistant strains by taking antibiotics until bacterial populations are extinct rather than just reduced; development of novel antibiotics

mans is caused by a virus descended from rinderpest viruses of cattle or their wild relatives that successfully shifted to life among humans. Tuberculosis in humans is caused by infection of the lungs with *Mycobacterium tuberculosis* or its close relatives, including *M. africanum* and *M. canetti,* which infect humans exclusively, and *M. bovis,* which can infect a diverse group of mammals, including cattle and humans. Recent analyses using large genomic data sets suggest that *M. tuberculosis* is the oldest of the group, although there are still many deaths caused by *M. bovis* passing from cows to humans.[18]

Influenza A virus infections in humans are generally the result of influenza viruses from waterfowl and other birds that recombine while residing in cells of domestic pigs before being passed on to humans. Recombination yields novel genetic variation that prevents development of lasting immunity in humans. Places where waterfowl, pigs, and humans live in close proximity provide a conducive environment for continual influenza A evolution and infection. In 1997, an influenza virus (strain H5N1) infected people directly from chickens, killing six of the eighteen people known to be infected. In a historical turnabout, molecular genetics studies enabled identification of the pathogen source within days, and drastic preventive measures were taken to prevent further human infections with this deadly strain. About 1.7 million chickens were destroyed in Hong Kong, where the infections occurred, and changes were made in the regulations for importation of chickens to Hong Kong. Future epidemics appear likely.

Similarly, some pathogens have passed from domestic animals to wild species, a phenomenon known as "spillover." Canine distemper virus from domestic dogs has recently infected wild African hunting dog *(Lycaon pictus)* populations, killing many and exacerbating problems for this threatened species. Canine distemper virus from domestic dogs has killed other African wildlife species as well, including silver-backed jackals, bat-eared foxes, and lions, among others. When Asian cattle breeds were introduced into Africa in 1889, their rinderpest viruses spread to wild bovids and some of their predators. Within a decade of Asian cattle rinderpest introduction, 90 percent of Kenya's wild cape buffalo population had been decimated. "Spillback" describes a situation in which a pathogen initially passed from domestic to wild species returns, potentially altered with newly evolved virulence for the domesticated species. For example, domestic cattle spread the bacterium causing brucellosis *(Brucella abortus)* to bison and elk in Yellowstone National Park, who, in turn, returned and spread the infection to different cattle populations. Even BSE or bovine spongiform encephalopathy, better known as mad cow disease, may be seen as resulting from domestication and associated agricultural practices. The initial outbreak was a result of humans eating beef contaminated with an infectious protein (a prion) picked up by cattle eating feed that included contaminated brain or nervous tissue of other cows.

Another unintended yet important consequence of maintaining domesticated animals is the effect of routinely giving antibiotics to livestock. When livestock are given antibiotics prophylactically to promote their growth by reducing bacterial infections, this can favor the survival of resistant strains of bacteria that can and do turn up in packaged meat at the grocery store.

Domestication in Evolutionary Perspective

Humans have been applying the basic concepts of evolution for millennia. By gathering and sowing seeds from preferred food plants and by promoting the successful reproduction of favored animals, we have been playing the role of natural selection agent. In many ways we have been masterful, facilitating our own successful colonization of the planet. By raising, preserving, and storing foods from domesticated plants and animals, we have dramatically increased the carrying capacity, or number of individuals that can be supported by the environment, across the globe. There are more humans alive now than there were during the entire 2 million years prior to the end of the last glaciation period 12,000 years ago. Food surpluses and increases in population sizes allowed development of human societies and skills that would not have been possible otherwise.

Simultaneously, we have wrought havoc on environments and the species that live or lived in them. Some of our water reservoirs are visible from the moon and few untrammeled areas remain. Most large, habitable tracts of land are tied up in agriculture or grazing of domestic plants and animals, or in urban areas supporting human populations who have little knowledge or experience in raising the domesticated species they rely on for food and clothing. Though we might claim otherwise, we much prefer domesticated species to wild ones. As has been noted by Steve Jones, there are about 50 million dogs in the United States, compared to about 10,000 wolves, which we kill and persecute because they kill a few of the planet's billions of sheep.[19]

Domesticated species have clearly evolved from wild progenitors using the same process of differential reproductive success that has driven and continues to drive the evolution of all life forms. Who has led this evolutionary transformation? The answer can only be "humans." The difference between evolution of domesticates and evolution of varieties in nature is the human role in domestication. Humans promote and ensure the reproductive success of individuals with desirable traits despite the fact that favored individuals might die on their own due to concomitant loss of adaptive traits or development of novel traits that would be deleterious in the wild.

Domestication is reasonably considered to be applied evolution. This application entails continual use of our growing understanding of biological knowledge, including evolutionary biological knowledge. This is not to claim that early agriculturalists conceived of common descent for all organisms, including humans, over the past 3.8 billion years. That would have been impossible until more recent times, when empirical evidence regarding evolutionary time spans and common descent for diverse organisms, not just look-alikes, became available and more widely appreciated. I do claim that humans have understood the basic premise of heritability (like father like son) and descent with modification among close relatives for at least 10,000 years. This pragmatic realization is the implicit assumption underlying all selective breeding efforts. Humans have used their keen powers of observation to detect small improvements in desired traits and to preferentially breed individuals having those traits, seeking to make those traits more common or even permanent among following generations. The archaeological evidence includes grains of increasing size over time, held within thinner coats; threshing tools, grinding tools, and seed storage vessels; dogs, cattle, and horses of changing proportion over time found in close association with human habitation; and, of course, range extensions for many selected species matching human migrations. The fact that humans have been applying evolution for thousands of years strongly implicates an early, intuitive understanding of the potential for change across generations.

Now that we have an understanding of some evolutionary mechanisms, can we ameliorate any of the problems resulting from artificial selection noted earlier? An evolutionary biologists' response, if not solution, to the problems of inbreeding is to promote outbreeding to increase genetic diversity. The benefits of outbreeding and subsequent genetic diversity are considered responsible, at least in part, for the success and preponderance of sexually reproducing vertebrate species compared to asexually reproducing species in nature. Populations under strong artificial selection, where only closely related individuals are mated, often suffer health problems without introduction of genetic material from distantly related individuals. The high market value of sperm or stud services of a star sire also promotes inbreeding, and the genetic and physical health of breeds suffers.

Most human cultures have restrictions or taboos against incest. These taboos may have arisen for biological reasons, attempting to reduce negative effects of inbreeding. Alternatively, some claim human incest taboos were implemented to avoid confusion about social roles within families, or both may have been involved. In any event, the biological consequences of inbreeding have long been known, and continued inbreeding in domestic animals is driven variously by desire for an aesthetically ideal form and by potential economic gain.

The bylaws of the AKC allow a dog to be registered as a legitimate representative of any particular breed only if its parents are also registered. This practice effectively prevents any introduction of new genetic variability that might alleviate genetic problems resulting from excessive inbreeding. Current judging standards are so constricting as to prevent any departure from the arbitrarily recognized "ideal" features. No program of screening individuals or their ancestors for genetic disorders and allowing breeding only among registered individuals can eliminate genetic defects, because recurrent mutation is a fact of life. In natural conditions, dispersal and other behaviors that promote genetic mixing also promote maintenance of genetic diversity and decrease the frequency of defective homozygotes. A better goal for breeders is healthy populations, rather than individuals conforming to inflexible standards and breed purity.

There is a role for genetic management in domestic species, as mapping of particular genetic diseases and development of assays for genes associated with particular diseases will allow informed choice of matings to avoid health problems associated with homozygosity for specific genes. Unfortunately, genetic diversity is already extremely low in some breeds, and matings across breeds may be the only recourse. There are kits for genetic screening of individual dogs and horses to see if they are carriers for various genetic disorders, but this is costly, labor intensive, and requires that breeders genotype their animals and not mate carriers to each other.

More important, especially for economically significant livestock and crops, is the maintenance of the genetic diversity present in less popular domestic breeds and their wild progenitors, if they still exist. Loss of breed diversity is a serious problem where commercial interests value one or a few breeds exclusively, and other breeds developed in earlier times, perhaps for different purposes, dwindle or go extinct. For example, many local cattle and chicken breeds developed in Europe and the Mediterranean region have been lost or have become threatened in recent times. Less popular breeds may serve to restore healthy genetic diversity lost from excessive inbreeding. Health problems associated with artificial selection provide further evidence for the relative efficiency of natural selection in honing viable populations.

In this sense, natural selection as imposed by humans will become increasingly important to human health and well-being. It will be useful in developing food, medicines, and raw materials for housing and clothes. Recent efforts in genetic engineering to alter genes or their effects in crops and livestock is a logical extension of our ancient efforts in selecting on organisms to increase desirable traits. The intent is the same, though the technology is radically different. We may similarly expect a mixture of positive and negative outcomes. As we learn the function of particular genes in particular taxa, we can begin experimenting to see how individual organisms succeed with those genes enhanced, blocked, or transplanted into individuals from species or breeds lacking them. We are a long way from wise application of genetic engineering; however, it seems inevitable that we will try, and that we will learn a great deal more about what is possible evolutionarily and what is not. Genetic engineering efforts have already inaugurated a new chapter in the 10,000-year chronicle of evolution in human hands.

3

EVOLUTION IN PUBLIC HEALTH AND MEDICINE

Despite our preferences, death and disease are with us still. From the standpoint of evolutionary biology, they are every bit as natural as life and good health. Through an understanding of evolution and its component theories, we may better understand how to combat disease and extend our life spans. The two revolutionary scientific hypotheses of germ theory and evolution first gained widespread acceptance in the latter half of the nineteenth century. The two ideas had not been brought to bear on each other until more recently, however. The study of microbes and of evolution have a mutually beneficial association. Increasing knowledge of one promotes understanding of the other. The concepts and methods of evolutionary biology are used widely to promote public health and in the practice of medicine.

The primary way in which germ theory benefits us is the ability to link specific pathogens to specific diseases, aiding preventive measures against infection. As noted in Chapter 1, Robert Koch outlined four criteria in the 1880s for linking particular microbes with particular diseases. The hypothesized pathogen should: be found in every case of the disease and be absent where the disease is absent; be grown in a pure form outside of the body; be shown to cause disease when administered to healthy individuals or cells; and be re-isolated from the experimentally infected cells or individuals and shown to induce disease in a new cycle of infection. These criteria are straightforward, but they can be difficult or even impossible to demonstrate in many cases. An evolutionary perspective helps us understand why this is so. For example, pathogens may cause disease in some individuals but not others because the host population has evolved variation in susceptibility, or the pathogen population has evolved variation in virulence, or both. Also, many pathogens, particularly viruses, are difficult to grow in cell culture in the lab due to uniquely evolved life history differences. Some pathogens, such as hepatitis B, hepatitis C, yellow fever virus, and HIV, have evolved to infect primarily humans, making it difficult to develop animal models for research. Frequent co-infection with multiple pathogens and multi-factorial causation can also complicate experiments seeking to link disease to specific microbes. These four postulates help frame the questions that must be asked in establishing disease causation, but evolutionary considerations impose some limits on their strict interpretation.

Understanding and coping with disease is a never-ending battle, but thanks to evolutionary theory we have several additional tools, which I will discuss in turn. These are: (1) understanding patterns of common ancestry (phylogeny) for pathogens and for hosts; (2) understanding evolutionary processes, especially natural selection, at work within populations of pathogens and hosts; (3) seeing the potential uses for natural selection as directed by humans on pathogens and on molecules; and (4) understanding human health problems in the context of human evolutionary history.

Applications of Phylogenetics

Phylogenetic trees show the genealogy of organisms based on comparisons of the traits of individuals. Branching diagrams indicate which species are closest relatives (sister taxa) among a larger set. Phylogenetic trees can also be used to infer the relative order of lineage divergence events and character changes in the evolution of organisms and their genomes. Phylogenies delineate the path of organismal change through time and provide the historical context for discovering its causes.

The methods for reconstructing phylogeny have a long and occasionally contentious history of their own. As with other analytical tools, phylogenetic methods are refined as our understanding of the assumptions inherent in different analyses improves and as our understanding of the nature of the data improves. If the traits of organisms changed at a constant rate and without redundancy (no convergence), phylogeny could be discovered with a simple metric tallying differences among groups of organisms. We now understand, however, that traits of all kinds, from behavior to molecular sequence, vary in their rates of change and that convergence can be common. There are, after all, only four nucleic acid bases possible for any DNA sequence position to be compared across taxa. Though there are many challenges for phylogeneticists, including some sets of taxa that defy direct comparison, great progress has been made.

Three general approaches are used to infer phylogeny: distances, parsimony, and likelihood analyses. Distance analyses consider measures of character similarity across taxa and place the most similar individuals or sets of individuals as closest relatives. Parsimony analyses focus on shared, derived traits by preferring the shortest tree that can be found, requiring the fewest assumptions of convergence in traits across taxa. Likelihood analyses use explicit models of variable rates of character change in optimizing the probability of hypothesized character changes, branch lengths, and phylogenetic relationships.[1]

The time scale and inclusiveness for phylogenies range from the tree of life, delineating relationships among all life forms spanning more than 3.5 billion years of history, to trees indicating relationships among individuals within populations whose divergences may have occurred only months or even hours ago, depending on the rates of reproduction. Recent growth of DNA databases, coupled with advances in phylogenetic methods, allow biologists to estimate patterns of relationships among bacteria and viruses previously inaccessible due to their small size and the lack of known traits available for comparison.

Phylogenetic analyses have several applications in understanding infectious diseases. A first application is identification of pathogens. We need to know what group of organisms or what taxon any particular pathogen belongs to. In practice, the relatively small number of common pathogens are identified by the symptoms they cause and rapid diagnostic tests that focus on presence or absence of one or a few unique traits. Nevertheless, accurate phylogenetic trees provide a comprehensive context for identification and are crucial for identifying rare or novel infectious agents, including those undergoing rapid evolution or colonizing new host species. Identification is not just a matter of deciding which known pathogen, as an unchanging "type," matches the unknown pathogen most closely. Rather, detailed identification is a matter of discovering a pathogen's ancestry. This stems from the fact that pathogens evolve, sometimes very quickly. Accurate identification of closest known relatives for new pathogens enables initial hypotheses about the pathogen's basic life history, as these traits tend to be shared among close relatives. This includes identification of likely host species, their geographic distribution and favored habitats, as well as the pathogen's likely mechanisms of reproduction and transmission. Often, where the limits of phylogenetic resolution are reached, analyses of variation within populations can provide valuable information for identification of pathogens.

A second way in which phylogenetic analyses help us understand infectious diseases is in the discovery of pathogens' history of transmission within and among host species. Knowing whether a human pathogen originated in nonhumans (zoonotic origins), and if so, which species provides the natural reservoir for the human pathogen, provides a background for understanding how we might reduce the pathogen's incidence and effects.[2] Third, phylogenetic analyses identifying closest relatives, together with population genetic analyses, can identify specific molecular differences associated with particular changes in pathogens, including traits such as virulence, rate of reproduction, transmissibility, and susceptibility to drugs, among others.

Phylogenetic trees and the study of population variation can also be used to ask questions about the factors influencing the spread of pathogens. Are particular pathogen clades associated with particular human behavioral, demographic, or geographic groups? In this way, phylogenies can tell us which groups of viruses, bacteria, or protozoans are most likely to give rise to new variants causing disease in humans, or which behavioral or demographic groups of humans are at greatest risk for various diseases.

Where past events can be dated, such as known first infection dates for individuals or populations, the potential exists to use those dates to calibrate average rates of pathogen sequence evolution. These rates can be used, in turn, to infer age estimates for other evolutionary events in other host populations or related pathogen lineages.

As noted above, phylogenetic trees delineate the course of evolutionary change through time, and provide the historical context for discovering the causes of that change. Phylogeny also provides the organizing pattern for delineating and classifying biodiversity. Let's briefly consider the phylogenetic diversity of pathogens and their variable natural histories. We can understand pathogens as the end-products of evolutionary history (as well as important factors in the evolution of the organisms they infect) because natural selection favors those traits enhancing reproductive success for both pathogens and their hosts. Seeing pathogen diversity from an evolutionary perspective can help us understand their potential responses to change in their environments. If we can predict how and when a pathogen will respond to environmental change, we may avoid their worst effects.

Phylogenetic Origins of Pathogens

Pathogens fall into three primary groups, given here in decreasing order of body size: protozoans, bacteria, and viruses. Protozoans, or protists, include amoebae, flagellates, slime molds, sporozoa, and algae—a diverse assembly of microscopic eukaryotes within the taxonomic domain Eukarya. Members of Eukarya are distinguished from members of the other two domains, Archaea and Bacteria, by two evolutionary innovations, a cell nucleus and a cytoskeleton. Approximately 115,000 species of protozoans are known, although this is a very rough index of their diversity. Many protozoans reproduce asexually, and it is likely that many distinctive lineages remain to be discovered.

Protozoans play important roles as both predators and prey

in the ecology of soil and aquatic habitats, and some have profound effects on human health in their role as specialized parasites. For example, malaria, a debilitating blood disease, is the result of infection by protozoa in the genus *Plasmodium.* African sleeping sickness and Chagas disease are caused by protozoan parasites in the genus *Trypanosoma.* For many AIDS patients the immediate cause of death is pneumonia caused by the lung-inhabiting protozoan fungus *Pneumocystis carinii.*

A glimpse of the phylogenetically disjunct positions of some of the worst human pathogens indicates the multiple, independent origins of disease-inducing parasites. Understanding phylogeny for this diverse set of organisms is a first step in understanding their various behaviors, favored habitats, mechanisms of reproduction, and susceptibility to antibiotics or other control measures, because these traits tend to be similar among closer relatives.

Bacteria, or prokaryotes, were the first form of life to evolve on the earth. Minute fossil bacteria date back about 3.5 billion years, and bacteria appear to have been the sole inhabitants of the planet for the next 2 billion years or so. In that time, bacteria pioneered the basic mechanisms of DNA replication and translation that are used by all the late-comers like ourselves. And, fortunately for us, they oxygenated the atmosphere too, largely through the activity of photosynthesizing cyanobacteria.

As recently as the 1960s, biologists thought bacteria to be a fairly uniform group. They believed that the majority of life's diversity was to be found among the plant and animal kingdoms. Carl Woese turned this view upside down beginning in the late 1960s by comparing ribosomal RNA gene sequences across bacteria and across all life forms. Some bacteria were found to be as different from each other as were any other organisms. Ultimately, Woese's phylogenetic analyses consistently identified three primary, early-diverging lineages in the

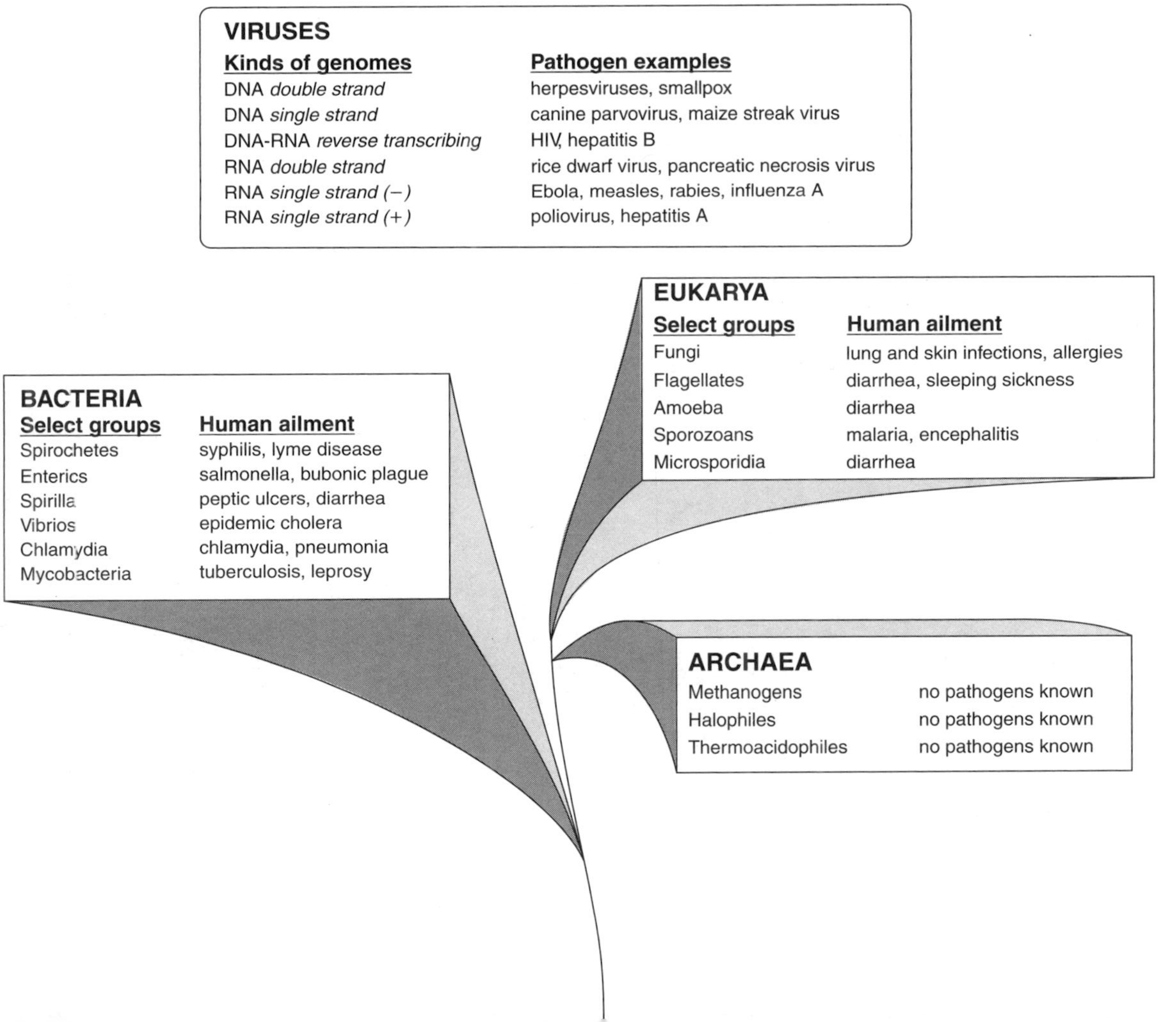

VIRUSES

Kinds of genomes	Pathogen examples
DNA *double strand*	herpesviruses, smallpox
DNA *single strand*	canine parvovirus, maize streak virus
DNA-RNA *reverse transcribing*	HIV, hepatitis B
RNA *double strand*	rice dwarf virus, pancreatic necrosis virus
RNA *single strand (−)*	Ebola, measles, rabies, influenza A
RNA *single strand (+)*	poliovirus, hepatitis A

BACTERIA

Select groups	Human ailment
Spirochetes	syphilis, lyme disease
Enterics	salmonella, bubonic plague
Spirilla	peptic ulcers, diarrhea
Vibrios	epidemic cholera
Chlamydia	chlamydia, pneumonia
Mycobacteria	tuberculosis, leprosy

EUKARYA

Select groups	Human ailment
Fungi	lung and skin infections, allergies
Flagellates	diarrhea, sleeping sickness
Amoeba	diarrhea
Sporozoans	malaria, encephalitis
Microsporidia	diarrhea

ARCHAEA

Methanogens	no pathogens known
Halophiles	no pathogens known
Thermoacidophiles	no pathogens known

tree of life. Two of them were bacteria (Bacteria, Archaea), and the third (Eukarya) included everything else.

Archaea includes producers of methane gas (methanogens), lovers of extreme saline environments (halophiles), and others adapted to near-boiling, sulfur-rich conditions (thermoacidophiles). To date, no archaeal pathogens are known. The taxon Bacteria includes all the other prokaryotes, from *Acetobacterium* to *Zavarzinia,* as well as the lineages whose descendants became mitochondria (from Proteobacteria) and chloroplasts (from Cyanobacteria). Although only about 4,000 species of bacteria (all kinds) have been formally recognized and named, this figure does not even begin to describe the abundance of different sorts. The biological species concept, restricted to sexually reproducing groups, has no meaning in describing fissioning bacteria, and the vast majority of habitats have not been surveyed.

Viruses are the smallest organisms on the planet, and consequently the last major kind to be described. On average, viruses are about 30 nanometers long (3×10^{-9} m) and 1/100 the size of most bacteria. Success in being small requires great economy in structure and content. For example, an HIV genome with a mere 10,000 bases of RNA and nine or so genes is only 0.0003 percent the size of the human genome.

Many viruses can produce thousands of offspring per hour in each of the hundreds or thousands of cells infected in a single host individual. This provides copious grist for the evolutionary mill, producing a multitude of winning virus forms and lifestyles that have succeeded in colonizing all other organ-

---

Figure 3.1 Representative pathogens from the three primary lineages of organisms (Bacteria, Archaea, Eukarya) and from viruses. Viruses appear to have arisen on multiple, different occasions from the genomes of other organisms and, possibly, independently of other organisms dating back to the origin of life.

isms, from bacteria to algae, fungi, plants, and animals, moving with them to all regions and habitats on the earth.

Getting a grip on viral diversity entails understanding how many different kinds of viruses there are and how they are related to each other. Early classification for viruses centered on the similarity in diseases or symptoms caused, the means of transmission, or the kinds of organisms or even body organs infected. For example, viruses able to induce swelling of the liver with accompanying fever and yellowing of the skin (jaundice) caused by buildup of a bile pigment were classified together as the hepatitis viruses. This included what are now seen as distantly related or unrelated groups such as hepatitis A virus, hepatitis B virus, yellow fever virus, and Rift Valley fever virus. Biochemical and molecular studies in the 1960s and early 1970s facilitated classification of viruses based on the nature of their genetic material, whether RNA or DNA, and whether the genome was double- or single-stranded. If the genome was single-stranded, the virus could be classified further depending on whether that strand was identical to the messenger RNA transcript (positive-stranded) or complementary to it (negative-stranded).[3] Beginning in the 1980s and 1990s, biologists sought to develop a taxonomy for viruses based on phylogenetic analyses of shared traits, primarily nucleic acid sequences. This is a work very much in progress with no guarantee of advance after the most obvious relationships are determined.

Based on similarity in the nature of the viral genome, strandedness (positive-stranded or negative-stranded) of the viral genome, capacity for reverse transcription, and polarity of the viral genome, six primary groups of viruses are generally recognized, comprising 73 families and 287 genera.[4] Fundamental differences among viral groups in the structure and function of their genomes support the view that viruses have arisen on multiple, independent occasions. In this view, viruses are not united in a single evolutionary clade, but instead represent a grade, sharing a convergently similar lifestyle.

How did viruses originate? Our understanding of ancient virus origins is extremely limited. They evolve rapidly and few characters, if any, are available for comparison between viruses and other organisms. Despite these severe limitations, two general hypotheses for the mechanism of viral origins have been identified. These rely on the same evolutionary mechanisms, including mutation, recombination, and natural selection, known to operate in more recent times and throughout the tree of life.

The primordial hypothesis holds that some RNA viruses have been present since the beginnings of life on the earth about 3.8 billion years ago. In this primordial hypothesis, simple RNA molecules, with strings of concatenated nucleotides, arose from pools of free nucleotides as a result of the chemical and physical attractions among singleton nucleotides. Simple RNA molecules now have been shown to be capable of copying themselves by serving as their own replication enzyme. The fact that viruses and their related genetic elements are ubiquitous within the cells or genomes of all life forms also suggests an early origin.

The escaped transcript hypothesis posits that viruses arose from messenger RNAs or other host-cell RNA or DNA molecules that acquired the ability to be replicated and packaged in a proteinaceous coat, enabling an escape from their cellular confines. Messenger RNAs routinely pass through the membrane of the nucleus on their way to the ribosomes in the cellular cytoplasm, where they are translated into amino acids. Successful passage through the nuclear membrane makes navigation of the cell wall seem feasible as well, though the mechanisms differ significantly. In this scenario, viruses evolved through a series of intermediate forms from an obligate intracellular progenitor.

There is a sense that a "significant fraction" of the primary kinds of viruses are now known, based on the low frequency for discovery of viruses that do not fit into existing families. The number of lower-level viral taxa that have been described, however, represent just the tip of the iceberg. Little survey work has been done for viruses outside of those infecting humans and our domestic animals and plants. We have no idea how many different viruses with unique capabilities infect archaebacteria, whales, slime molds, or other forms of life. The value of increased surveys and phylogenetic analyses of virus diversity among nondomesticated species is revealed each time we encounter a novel viral pathogen and find that we have little or no understanding of its natural host range or means of reproduction.

Pathogen Case Histories: An Evolutionary View

West Nile Virus

On 23 August 1999 two cases of encephalitis were reported to the New York City Department of Health. A few days later, six more cases of encephalitis were identified, all located in the same two-mile square area in Queens. An infectious agent was suspected, but what was it? An initial screening indicated the presence of antibodies for St. Louis encephalitis virus, an arbovirus in the genus *Flavivirus,* in the blood of the patients. Antibody tests do not necessarily provide definitive identification, however. A single pathogen may elicit multiple kinds of antibodies, and more than one pathogen may elicit the same antibody response. At the same time the encephalitis cases were reported, exotic birds were dying at the Bronx Zoo, and wild crows in the surrounding area were dying as well. Autopsy and lab tests on the dead birds revealed encephalitis and an unrecognized virus.

Eventually, phylogenetic analyses of viral genes identified the human virus as a West Nile virus (WNV) previously known only in Old World countries. The phylogenetic tree indicated that the human isolate was most closely related to a WNV isolated from a dead goose in Israel in 1998.[5]

This and subsequent phylogenetic analyses indicated the unwitting role of migrating birds in the spread of WNV. Phylogenetic identification of WNV in humans and birds allowed public health officials to identify the birds infected with WNV as the reservoir and mosquitoes as the carrier for the human WNV cases. This, in turn, led to the dissemination of information by public health officials, who urged the public to avoid mosquito bites. Phylogenetic analyses of portions of the nucleocapsid and envelope protein genes indicate rapid evolution of distinctive geographic clades of the virus in the United States, and this information will enable biologists to follow the geographic movement of variant WNV strains and their genomic evolution as they change and adapt in North America.[6]

Phylogenetic analyses of WNV and its relatives in the family Flaviviridae, such as dengue fever virus and yellow fever virus, also provide some insight into understanding the recent success and increasing prevalence of this group. Flaviviridae includes two primary clades; one clade includes viruses carried by ticks, the other includes viruses carried mostly by mosquitoes. Knowing to which clade any given flavivirus belongs is important in designating control measures. Further, increasing human populations over the past two hundred years appear to have facilitated diversification and increase in the numbers of viruses in the clade carried by mosquitoes.[7] This indicates, again, the need to consider precautions against mosquitoes.

Knowledge of WNV phylogenetic relationships can help us identify potentially useful existing vaccines and provide a starting point for developing new ones. Attempts to develop vaccines for WNV are beginning with existing vaccines licensed for immunization against yellow fever, caused by another flavivirus. For the WNV vaccine, biologists are attempting to substitute a surface protein from WNV for the same gene from the yellow fever virus vaccine. This phylogenetically

Figure 3.2 Evolution of West Nile viruses.

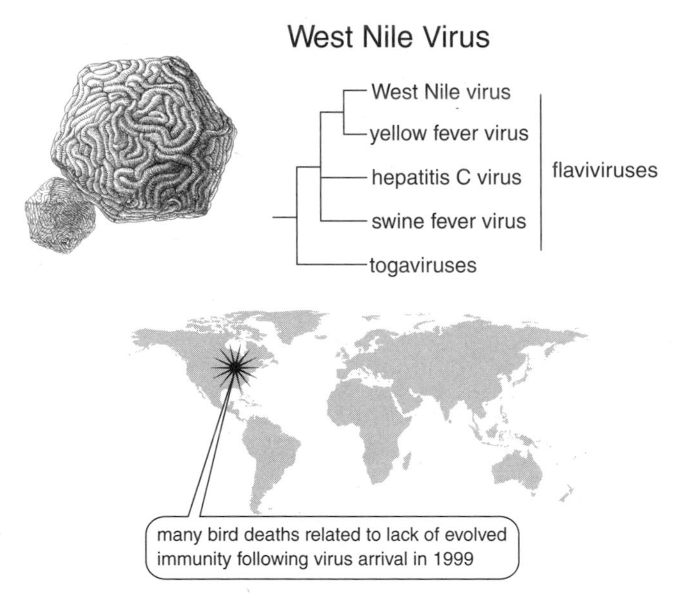

Evolution & Adaptation

- ***Evolutionary history:*** related to other viruses in the + single-stranded RNA virus family Flaviviridae
- ***Adapted to:*** live in mosquito salivary glands and reproduce in the blood of birds, with humans, horses, dogs, and other animals as incidental hosts
- ***Dispersal:*** transmitted among birds by mosquitoes
- ***Recent evolution:*** new strains evolving in North America, following arrival in New York from the Middle East in 1999; virus spread westward, reaching California in 2002
- ***Minimizing pathogen success in humans:*** reduce mosquito populations and their opportunities for contact with humans

informed approach of creating chimeric vaccines is being used to develop vaccines for other flaviviruses such as dengue fever virus and Japanese encephalitis virus.

Marburg and Ebola Viruses

In summer of 1967 three employees at a vaccine-producing facility in Marburg, Germany, developed flu symptoms. These flu symptoms led to rashes, diarrhea, vomiting, massive hemorrhaging, and after several weeks, death. Similar outbreaks occurred at about the same time among live animal handlers and health care workers in Frankfurt, Germany, and Belgrade, Yugoslavia, with similar results. The virus responsible for these deaths was isolated and characterized. Until that time, it was unknown in humans. The virus was determined to have been picked up through human contact with blood, kidney tissue, or cell cultures from African green monkeys imported from Uganda that harbored the virus.

Nine years later, in 1976, a deadly epidemic with symptoms eerily similar to those caused by Marburg virus swept through remote villages in Sudan and Zaire. The disease was thought to be yellow fever or typhoid at first. Once isolated and scrutinized, however, this virus also turned out to be previously unknown and was phylogenetically closely related to the Marburg virus. This new virus was named Ebola after a river near the site of the initial outbreak.

Phylogenetic analyses of RNA sequences for both Marburg and Ebola along with comparisons of their genome structure were valuable in identifying these unknown viruses as close relatives, sharing common ancestry with Paramyxoviridae viruses (including measles and mumps viruses) and Rhabdoviridae viruses (including rabies viruses). These two families plus the new family Filoviridae, which includes the Marburg and Ebola

Figure 3.3 Evolution of Marburg and Ebola viruses.

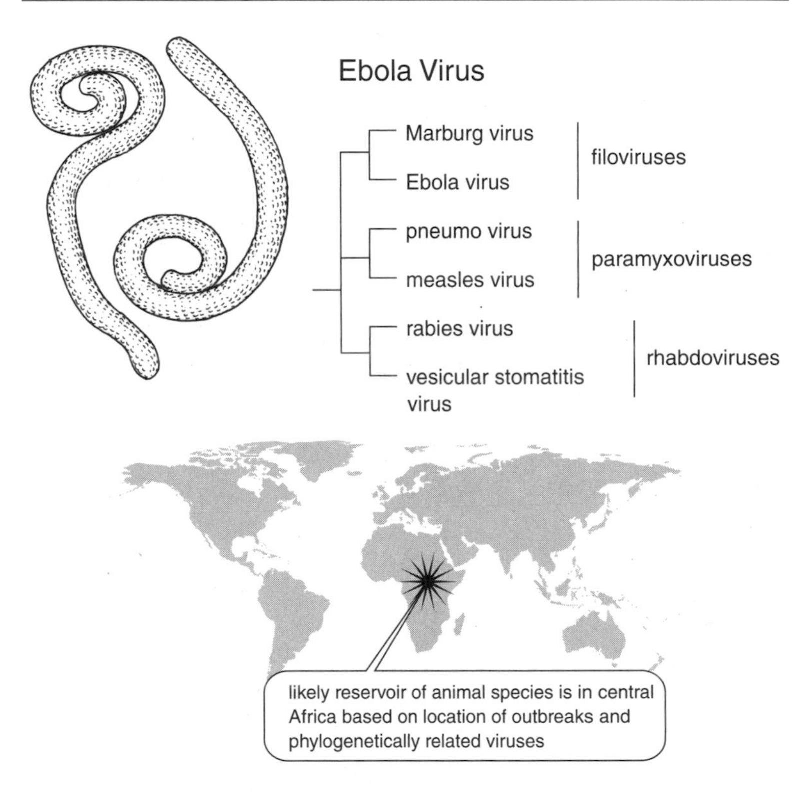

Evolution & Adaptation

- ***Evolutionary history:*** related to other single-stranded RNA viruses in the family filoviridae
- ***Adaptations:*** poorly known; at present poorly adapted for long-term survival in human populations
- ***Dispersal:*** transmission to humans by contact with infected mammals and spread among humans by contact with secretions from those already infected
- ***Recent evolution:*** Four identified evolutionary variants from Zaire, Sudan, and Ivory Coast have caused disease in humans; Ebola-Reston virus has caused disease in nonhuman primates.
- ***Minimizing pathogen success:*** reduce contact with potentially infected animals; isolate infected individuals

viruses, comprise one of the few described viral orders, the Mononegavirales.

Subsequent outbreaks of Ebola have occurred in Gabon, Zaire, Republic of the Congo, and Uganda, with over a thousand known human fatalities to date[8] and an unknown but alarming number of deaths among gorillas in the Republic of the Congo. The natural host species for this apparent zoonotic disease remains to be determined, though pieces of its RNA genome have been found, sequenced, and phylogenetically identified in rodents in the Central African Republic.[9] This illustrates the need for further surveys of potential hosts for Ebola virus to provide isolates and sequences for phylogenetic analyses aiming to track zoonotic sources and transmission histories. Until the natural reservoir and transmission steps are determined, Ebola will continue to be a fearsome disease.

Influenza A Virus

Influenza has a long and deadly history. Epidemics of highly contagious respiratory disease sweeping through Athens in 430 BCE, laying waste to Charlemagne's army, decimating regions of Europe and Africa in 1580, and killing thousands in the Caribbean and eastern North America in 1647 have all been linked to influenza. Perhaps the best-known pandemic is that of 1918, in which 20 to 40 million people were killed worldwide in less than two years. The name "influenza" originated in the early 1400s in Italy, where epidemics of the disease were thought to be under the influence of the stars, for lack of any better explanation. During the late nineteenth century, the bacterium *Haemophilus influenzae* was thought to cause influenza because it was a common throat infection in many influenza patients. Evidence for a viral cause was not found until the 1920s, when a disease similar to human influenza was demonstrated to be transmissible among pigs following experi-

Figure 3.4 Evolution of Influenza A viruses. Phylogeny is shown for the NP gene rather than for the entire genome on account of frequent mixing (reassortment) of genome components among virus strains.

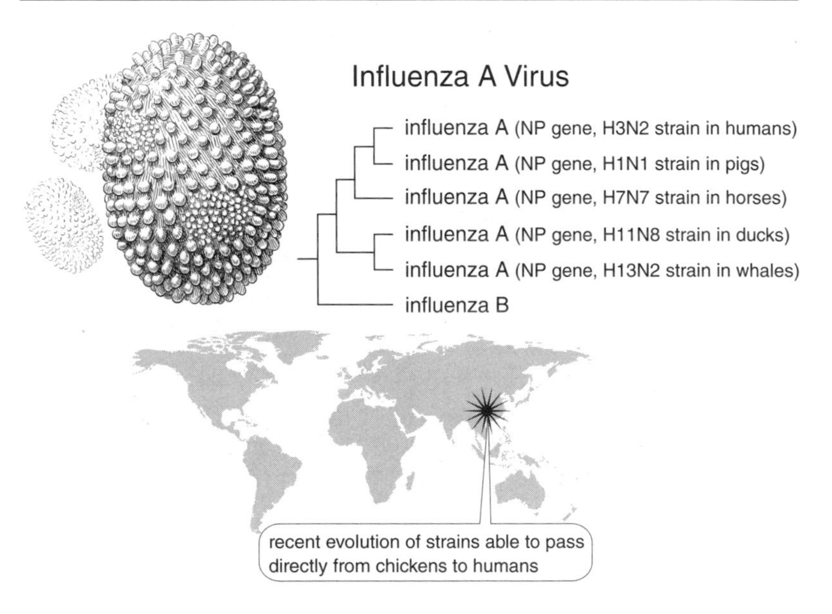

Evolution & Adaptation

- ***Evolutionary history:*** shares common ancestry with other single-stranded RNA viruses in the family Orthomyxoviridae
- ***Adapted to:*** continual rapid change as a result of frequent shuffling of genome segments and natural selection; coexistence with healthy populations of bird and mammal hosts
- ***Dispersal:*** transmission to and among humans by breathing contaminated air
- ***Recent evolution:*** frequent origin of novel forms requires continual vaccine development; evolution of new, virulent strains have potential to elude existing host defenses and become fatal as seen in epidemics of 1918, 1957, and 1968, and smaller-scale recent outbreaks with some strains able to pass directly from birds to humans
- ***Minimizing pathogen success in humans:*** enhance surveillance, minimize opportunities for virus mixing and transmission among host species, and development of vaccines

mental infection with filtered, viral material. The vast reservoir of influenza viruses in other vertebrates, especially birds, was not discovered until the 1970s.

Phylogenetic analyses of influenza A viruses beginning in the late 1980s indicate that wild birds are the zoonotic reservoir for human influenza A, and evidence continues to accumulate that domestic pigs are often intermediary hosts between birds and humans. This is based on the observation that influenza A gene phylogenies show pig and human viral gene sequences as closest relatives that are recently evolved from avian viral gene lineages. Wild bird populations, mostly waterfowl and gulls, harbor the greatest diversity of influenza A virus types and show little evidence of ill health when infected. Both of these observations are consistent with a longer evolutionary history for influenza viruses in birds than in pigs and humans, in whom reduced virus variation is found and acute disease appears to be more common. Though there are exceptions, greater diversity often indicates older divergences, and greater disease often indicates a younger, less stable association.

Influenza A viruses have genomes comprised of eight unique segments of single-stranded RNA. When single host cells are infected with multiple, different influenza A virus strains, these segments can be mixed and matched in the viral offspring. This form of recombination, known as "reassortment" or "shift," provides the near-kaleidoscopic variation that enables reconfigured influenza A strains to overcome the immunity developed in humans to the more recently circulating strains with a different configuration of the genomic segments. Glycoproteins of hemagglutinin (H) and neuraminidase (N) on the surface of the virus are particularly important in helping the virus avoid attack by the host's immune system. H and N genes of multiple types exist and their reassortment results in a virus that can cause an epidemic. For example, the 1918 epidemic was caused by H_1N_1 type viruses, the 1957 epidemic (Asian flu) by H_2N_2, the 1968 epidemic (Hong Kong flu) by H_3N_2, and the 1977 epidemic (Russian flu) by new recurrence of H_1N_1. Most recently, H_5N_1 influenza A strains have jumped from chickens or waterfowl to humans on multiple occasions, resulting in over 55 deaths between 2003 and mid-2005. So far, there are no known incidences of human-to-human transmission for these lethal H_5N_1 strains. Their lethality and the ease of human-to-human transmission for related influenza A strains points to the very serious threat posed by this rapidly evolving virus. Just as in identifying zoonotic sources, the different origins and evolutionary histories for strains and their RNA segments is elucidated in greatest detail with phylogenetic analyses.

Surveys and comparison of influenza viruses in birds reveal a great deal of natural variation, indicating that virus-carrying birds will always present a potential for human pandemics. Health officials have recommended that domestic birds and pigs (the apparent intermediary between humans and the avian zoonotic reservoir) be kept in enclosed facilities to prevent contact with wild birds.

Influenza A evolution also proceeds by mutation and substitution of individual amino acids, which can have significant functional effects. Comparative analyses revealed that a single amino acid change in the hemagglutinin gene converted a H_5N_2 virus with low virulence in chickens to one with high virulence in chickens, prompting a poultry epidemic that began in Pennsylvania in 1983.[10] Phylogenetic analyses facilitate linking of specific genomic changes with changes in virulence. For example, phylogenetic analyses of influenza A virus H genes, mapping specific amino acid changes to their position of occurrence on the tree, showed that eighteen amino acid positions at antigenically important sites were naturally selected for adaptive change, enabling escape from hosts' immune surveillance. Further, influenza lineages experiencing the most change at those antigenically important sites also had the greatest survival rates.[11] This information about which lineages have frequent, recent adaptive amino acid change at antigenic sites may help public health officials select appropriate influenza A strains for use in the continual development of vaccines.

Thus, evolutionary concepts applied to the study of influenza A help explain the continual generation of new and potentially lethal strains. Recombinant strains are generated when genome segments are randomly reassorted in new, individual viruses, and those variant viruses are naturally selected for survival and successful reproduction in hosts who are immune to recent but not current recombinant viruses. Following the evolutionary path of these recombinants helps us detect which of their protein structures and which amino acid positions are more likely to allow escape from the hosts' immune defenses. This knowledge helps us predict which lineages are most likely to survive and become a threat.

HIV

Acquired immunodeficiency syndrome (AIDS), resulting from infection with human immunodeficiency virus (HIV), was first recognized in the early 1980s. After a twenty-year rampage, more than 16 million people have died and over 42 million are infected worldwide. Mortality rates are declining where there is access to powerful antiviral drugs, which can cost $55 a day or more. Mortality rates continue to increase in populations without access to antiviral drugs, especially in Africa and parts of Asia, and infection rates continue to increase globally, unabated. At present over 80 percent of all AIDS deaths occur in sub-Saharan Africa, and AIDS is now the leading infectious cause of death worldwide, having recently surpassed malaria in its reach.

The pandemic could have been much less severe had HIV been contained within regional human populations when it first emerged. We were unprepared, and remain unprepared in many ways, to handle such a rapidly evolving, sexually transmitted disease. The pandemic could also have been much worse, however. The HIV pandemic coincided with the advent of new molecular techniques. HIV, a retrovirus, is one of the most comprehensively studied organisms. It is most commonly picked up during unprotected sexual contact. HIVs preferentially infect white blood cells, known as CD4+ cells, which play an important role in activating the immune system. Once inside, HIV particles remain hidden from other immune-system defenses and co-opt the CD4+ cellular machinery to assist in making copies of the HIV genome. The HIV particles then insert the HIV genome into the host's genome, where it regulates the production of new HIV particles. As part of HIV's success in reproducing, the CD4+ cell is destroyed as the HIV progeny emerge. On average, the immune system can replace 10 billion CD4+ cells a day, but years of balancing loss with replenishment takes its toll and the eventual shortage of CD4+ cells leads to the suite of symptoms known as AIDS.

Some antiviral drugs successfully block the HIV reverse transcriptase gene from working to replicate its own genome; others block activity of the virally encoded protease gene, which helps assemble the new HIV particles. None of these medications with protein site-specific effects could have been developed with pre-1980s technology, analyses, or understanding, including insights derived from comparative, evolutionary analyses of HIV lineages differing in their degree of virulence.

Phylogenetic analyses reveal two primary lineages of HIV. HIV-1 is the primary cause of the virulent pandemic, and HIV-

Figure 3.5 Evolution of human immunodeficiency viruses.

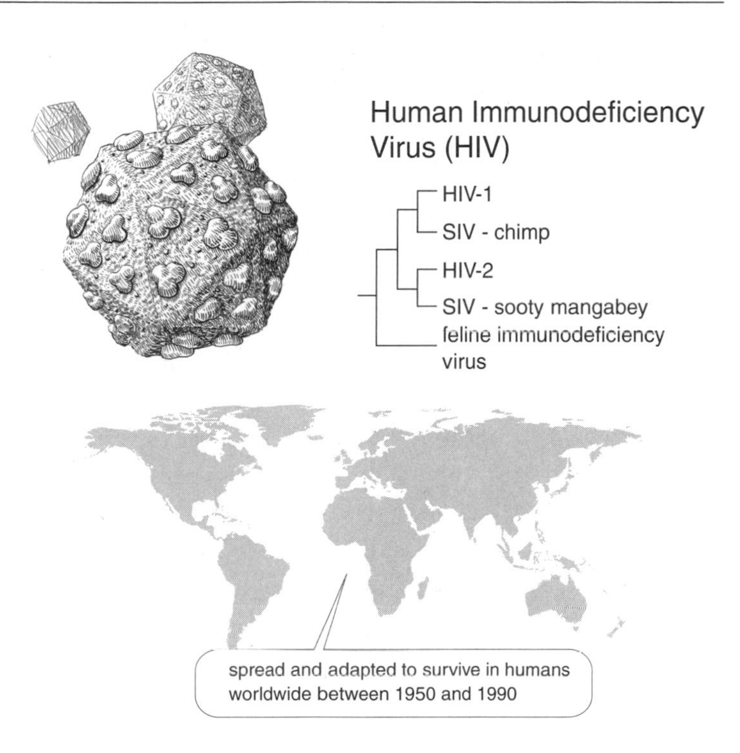

Evolution & Adaptation

- ***Evolutionary history:*** shares common ancestry with immunodeficiency viruses infecting other primates and other mammalian orders
- ***Adapted to:*** deceive hosts' immune systems by appearing as "self"; high levels of reproduction, and potential dispersal, prior to causing disease
- ***Dispersal:*** sexual transmission among human hosts
- ***Recent evolution:*** at least two independent colonization events by viruses (for HIV-1 and HIV-2) moving to humans from nonhuman primates; adaptation and dispersal from equatorial Africa to all other human populations during the latter twentieth century; rapid and repeated evolution of strains resistant to antiviral medicines
- ***Minimizing pathogen success in humans:*** avoid direct contact with potentially infected other primate species; reduce selection favoring virulent strains within humans by reducing transmission via unprotected sexual contact

2 is less widespread and less virulent. HIV-1 is a zoonotic infection passed from chimpanzees in western equatorial Africa, where we know it as SIVcpz (simian immunodeficiency virus), to humans sometime during the first half of the twentieth century. HIV-2 is a second, independent zoonotic infection of a different SIV passed from another primate, possibly sooty mangabeys, to humans during the same general time period.[12] Over twenty different primate immunodeficiency viruses have been described to date, and although the frequency and relative timing of host-species shifts is not known, it is clear, as indicated by phylogenetic analyses, that transmission between species can and does occur. It is quite possible that local epidemics caused by immunodeficiency viruses passing among primate species (including humans) have occurred periodically in the past and have gone unrecorded. Infection can readily occur during butchering and eating of wild animals or during handling of live captured animals. Phylogenetic analyses inform us directly about the importance of avoiding contact with blood or other potentially infected tissues from other primates, particularly chimpanzees harboring closely related SIVs.

Phylogenetic analyses show that the same general processes of colonization by a small number of founders with subsequent diversification and adaptation that is characteristic of many animal "species radiations," such as cichlid fishes in Lake Malawi and finches in the Galapagos Islands, operate in the evolution of HIV-1 varieties, both regionally and globally.[13] HIV-1 subtypes that are based on phylogenetic relatedness were recognized during the 1990s and tracked as they spread across and among continents. Origin and differentiation among Lake Malawi's 400 or so cichlid fish species required 2 to 3 million years, and is considered a rapid species radiation by vertebrate standards. By comparison, divergences among and within successful HIV-1 subtypes have occurred in a matter of years or even months.

Recombination among HIV lineages results from template switching during replication when multiple HIV particles coinfect single host cells. When replication begins, with one particle as a template but then switches to another, this can yield recombinants among distantly related HIV-1 lineages. Recombination is increasingly recognized as important in the evolutionary history of HIV. Factors such as frequent recombination and recolonization make HIV-1 history complex, and many events are or will become unrecoverable as time passes. However, careful phylogenetic analyses remain the best method for attempting to elucidate those events.

Rapid rates of HIV sequence change, driven largely by low-fidelity replication, enable use of phylogenetic analyses to help identify transmission sources and routes within and among human populations. In turn, this enables education and prevention efforts aimed at minimizing future infection rates. This potential to reveal transmission routes has been demonstrated in analyses recovering a known transmission history, in which a man infected with HIV in Haiti passed the virus to several women in Sweden in a known sequence.[14] Phylogenetic analyses have been used to identify explicit sources and networks of transmission, for example, among sets of people with high-risk behaviors—such as male homosexuals and intravenous drug users—as well as forensic cases involving a dentist in Florida who infected several of his patients, a physician in Louisiana who intentionally infected a nurse (see Chapter 6), and a nurse who infected a patient.[15]

Detailed phylogeny for HIV-1 taxa helps in determining which sequence-level changes are associated with traits such as resistance to antiviral medicines, and which particular sequence sites are subject to accelerated rates of change due to selection pressure imposed by hosts' immune systems. Resistance to various antiviral medications can be conferred by one or a few amino acid changes, and lab experiments as well as genotyping and phylogenetic analyses of circulating HIV-1 strains have shown that the same mutations conferring resistance arise repeatedly and can be maintained by selection.[16] Rapid and recurrent evolution of resistance points to the value of having multiple antiviral medications targeting a variety of HIV sequence sites and using them in combination.

Additional evolutionary insight that might ultimately enhance therapy involves the study of natural genetic variation providing immunity. Successful entry of HIV into human target cells requires interaction with co-receptors, and a 32 base pair deletion in the CCR5 human co-receptor gene has been found to provide resistance to HIV infection. Reconstruction of the genealogy of this mutation and estimates of its frequency among populations have been used, together with population genetics (coalescence) methods, to estimate its date of origin. This was determined to be the late thirteenth or fourteenth century. It has been suggested that this mutation might have been selected for at that time because it provided a similar resistance to infection by the bacterium *Yersinia pestis,* the agent of bubonic plague.[17] An evolutionary approach focusing on natural variation within populations and genealogy has shown why some individuals do not become infected by HIV despite repeated exposure. This new knowledge about the origin and mechanism providing resistance to multiple pathogens will certainly be useful in future work to reduce infection rates.

Anthrax

Written descriptions of human and livestock plagues with symptoms very similar to those of anthrax date back to the Siege of Troy (about 1200 BCE) and the writings of Homer soon after. The disease has likely been an important culling agent in wild herbivorous mammals at times of overpopulation throughout history. Anthrax was transformed from a common, though intermittent, scourge to a less widespread problem by the development of an effective livestock vaccine in 1937. However, anthrax continues to be an endemic disease in parts of sub-Saharan Africa, Turkey, and southeast Asia, and the hardiness and pathogenicity of its spores make it appealing to terrorists.

Anthrax is an acute infectious disease of grazing mammals caused by the bacterium *Bacillus anthracis.* Herbivorous mammals, including cows and sheep, tend to pick it up by ingesting contaminated soils along with their food, and humans are infected, in turn, by contact with meat, hides, or hair from infected animals. *B. anthracis* is a rod-shaped, nonmotile bacteria that forms spores (dormant individuals) when exposed to oxygen. These spores are famously tough—capable of lying dormant in the ground for decades until picked up by an unsuspecting host, in which they then reproduce (usually in lymph nodes and spleen) and manufacture toxins. The *B. anthracis* life cycle is continued as the infected host develops high fever, inflammations, bleeding from orifices, and dies, often within a few days of the initial infection. The bloody exudate and decomposition of the carcass liberate the next generation of bacteria to sporulate and await another unsuspecting host.

Routes of anthrax transmission are not known completely, and phylogenetic analyses are a potential means for their discovery. Phylogenetic analyses require variable traits, and *B. anthracis* strains tend to be unusually similar in their molecular sequence, making the task difficult. However, recent analyses focusing on highly variable DNA markers within the genome have identified at least six genetically distinct groups, with the greatest phylogenetic diversity occurring among African isolates.[18] This is consistent with an African origin for *B.*

anthracis. Further, the high levels of similarity among strains indicate possible recency of origin or the occurrence of one or more recent reductions in population size or both. More de-

Figure 3.6 Evolution of bacteria causing anthrax. Phylogeny shows three of multiple *Bacillus anthracis* strains distinguishable using rapidly evolving DNA markers.

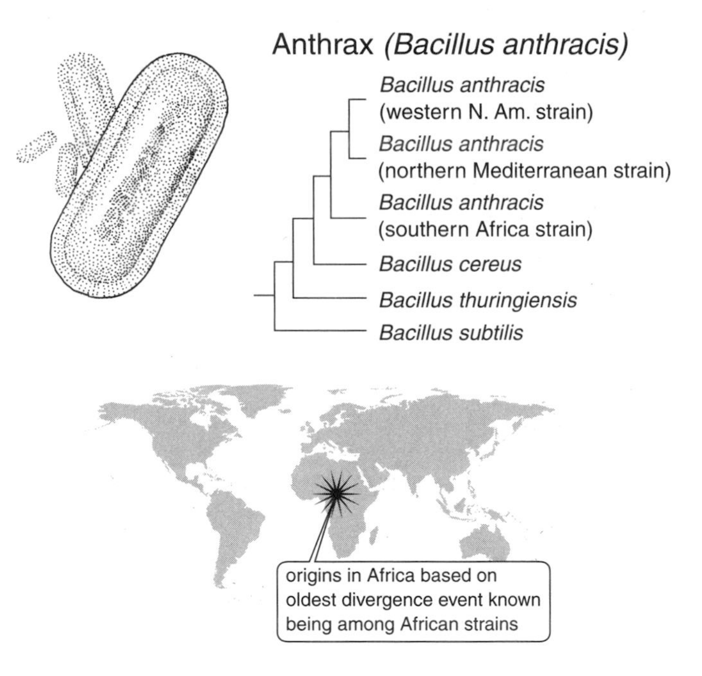

Evolution & Adaptation

- ***Evolutionary history:*** closely related to *Bacillus thuringiensis,* which parasitizes insects and is used for pest control, and to *B. subtilis,* which engages in cannibalism
- ***Adapted to:*** long periods of dormancy in soil; produce toxins weakening or killing hosts while reproducing
- ***Dispersal:*** by herbivores ingesting soil and consumers eating contaminated meat
- ***Recent evolution:*** some strains with antibiotic resistance
- ***Minimizing pathogen success in humans:*** use caution in handling infected animals or carcasses; enhanced surveillance and vaccine development; treatment with antibiotics

tailed sampling of *B. anthracis* is likely to provide better resolution.

The reason why *B. anthracis,* a soil-dwelling bacterium, makes toxins proving fatal to hosts in so many instances also remains to be fully explained, and again, a phylogenetic approach provides some insight. Until recently, distinguishing *B. anthracis* from its apparent closest relatives, including *B. cereus,* has been difficult. There are some differences in the presence of plasmid-born toxin genes between *B. anthracis* and *B. cereus,* but analyses of known chromosomal genes showed relatively little variation. However, analyses of newly sequenced whole chromosomal genomes reveal much greater variation distinguishing *B. anthracis* and *B. cereus.* Whole-genome data also show that the bacteria share a set of genes indicating a preference for proteins and complex carbohydrates as nutrients, rather than the simple sugars and carbohydrates of plant origin as found in related soil-dwelling bacteria.[19] This same food preference is also found in *B. thuringiensis,* an insect pathogen used as a pesticide, which turns out to be the closest known relative to the *B. anthracis* and *B. cereus* clade. This suggests that the most recent common ancestor of *B. anthracis* and *B. cereus* was not a harmless soil-dwelling organism, as had been thought by some, but was instead a predator of sorts feeding on animals, living or dead. If true, this phylogenetic perspective helps explain the existence of *B. anthracis*' toxic armory as an evolutionary legacy, providing better context for studies of virulence and its regulation.

Use means potential abuse, and it bears repeating that phylogenetic trees are hypotheses about relationships and not proof. Phylogenetic analyses suffer when the characters used are poorly suited, having excessively high or low levels of variation. For example, if the characters' rate of change is so high that convergent similarity is common, analyses can be misleading, unless the issue of convergence is addressed explicitly. Recombination among lineages, which is common among many virus groups, and skewness in the relative frequency of different nucleic or amino acids in molecular sequences analyzed present additional difficulties. The thorough sampling of individuals is also an important prerequisite to successful phylogenetic analyses, and inadequate sampling can contribute to errors.

The Role of Infectious Agents in Chronic Disease

Proving the cause-and-effect relationship between particular pathogens and particular infectious diseases has been one of science's great achievements. For example, knowing that malaria is caused by protozoans in the genus *Plasmodium* transmitted by mosquitoes and that yellow fever is caused by a flavivirus also transmitted by mosquitoes facilitates preventive measures and the saving of human lives. Most of the pathogens causing acute human diseases have already been identified, and novel diseases, such as Ebola hemorrhagic fever, usually can be ascribed quickly to particular pathogens. However, these are *acute* diseases, characterized by rapid onset and rapid progression with severe symptoms.

There are many chronic diseases, characterized by slower onset and slower progression, whose causes are less well understood. Multiple, noninfectious factors such as human genetics, diet, and environmental toxins or stress contribute to causation. Potential contributions of pathogens to chronic diseases are difficult to recognize because the symptoms take longer to develop, are more subtle, and often work in concert with genetic mutations and environmental variables. All of these issues make the source of chronic diseases much more difficult to establish.

There are chronic diseases now known to be infectious in origin. From the 1950s through the early 1990s peptic and duodenal ulcers (chronic sores on the inner lining of the stomach and small intestine) were attributed to stress, diet, and genetic predisposition. These ulcers are now understood to stem most often from infection with the bacterium *Helicobacter pylori.* The infectious component was missed for several reasons, including previous difficulty in determining presence of the bacteria in the stomach (or antibodies to them), the fact that only a small proportion of those infected with *H. pylori* actually develop ulcers (there are other risk factors involved, such as alcohol use), and the relatively cryptic form of bacterial transmission among humans via contact with fecal material. Further, Robert Koch's postulates for identifying pathogens responsible for acute diseases (discussed at the beginning of this chapter) are less successfully applied in studies of chronic disease, even if there is an infectious component, because of the disease's delayed onset and intermittent nature.

Other examples of chronic diseases with infectious components are tuberculosis, caused by infection with the slow-growing bacteria *Mycobacterium tuberculosis,* and various sexually transmitted diseases such as syphilis, caused by infection with the bacteria *Treponema pallidum.* These diseases include an initial acute phase, suggesting infectious causation, which later grades into a chronic condition, making their pathogenic source more obvious. Chronic rheumatic fever, characterized by fever and joint inflammation, was linked to prior infection with the bacterium *Streptococcus pyogenes,* which causes acute strep throat, after a half-century of debate. Cancers, characterized by abnormal cell growth and proliferation, have multiple causes, and infectious components have been and are being established for some. Peyton Rous first demonstrated infectious cancer causation in chickens in 1909 via what is now known as Rous sarcoma virus (though Rous could only infer the presence of an unseen infectious agent). Papillomaviruses are implicated in human cervical cancer, as are hepatitis B and C viruses in liver cancer. HTLV-1 is associated with white blood cell (T-cell) leukemia, though not all infected individuals develop the disease, and the latency period can be fifty or more years.

Awareness of the potential involvement of pathogens in debilitating chronic diseases has grown, and some have suggested that many if not most unexplained chronic diseases will eventually be found to have infectious components.[20] One of the reasons given in support of this is that any genetic mutations in humans inducing debilitating chronic disease would be selected against and reduced or eliminated from populations. However, if the pathogen's effects do not negatively affect reproduction, are not manifest until after reproduction, or if there are beneficial effects prior to reproduction, selection will be unable to act on underlying genetic causes directly.

Is there a role for evolutionary biology in assessing possible infectious components in chronic diseases? Yes, but this role is not well developed and is likely to be only supportive. There are evolutionary predictions, as indicated in my discussion of acute diseases in this chapter, that can be considered in testing hypotheses of infectious causation for chronic diseases. Yet they will be difficult to test or implement because of the delayed onset and cryptic nature of such diseases. Accurate phylogenies for infectious pathogens should reflect their history of dispersal (transmission) and subsequent diversification. If transmission is by direct contact among susceptible and relatively sedentary hosts, phylogenies for pathogens will track physical proximity of those host individuals. For example, this general pattern has been seen in HIV-1 isolates, especially early in the pandemic, which clustered geographically. This correspondence between phylogenetic and geographic proximity will break down to the extent that individuals or transmission vector species migrate to distant populations. For this approach to be applied, there would have to be detectable disease pathology among individuals as well, and this will be problematic for chronic diseases. In the long term, however, with eventual expression of latent disease, the trend should hold.

Evolutionary biology includes the study of organismal ecology and life history traits, which may also contribute to an assessment of potentially infectious diseases. Some pathogens are seasonal in their occurrence as a result of environmental needs or seasonal activity patterns for their vectors. The involvement of those vectors in disease may be implicated where chronic disease has a seasonal component. For example, among acute diseases discussed, WNV is spread by mosquitoes, and transmission and WNV disease are closely associated with mosquito reproductive activity and dispersal in late summer.

There are a few instances in which this approach might be applied to chronic disease. Some evidence links breast cancer to infection with Epstein-Barr virus, human papillomaviruses, or a human homologue of mouse mammary tumor virus (HHMMTV). These are given as co-factors for the disease along with diet, hormone metabolism, and genetic factors.[21] Similarly, there is some evidence linking *Chlamydia pneumoniae* infection to atherosclerosis (arterial clogging), though there is also growing evidence for genetic and environmental causation as well. There is speculation that HHV-6 and *Chlamydia pneumoniae* are linked to multiple sclerosis.[22] The definitive evidence will come as a result of observing mechanisms of pathogenesis and linking them to specific pathogens. Nevertheless, determining close phylogenetic relationships for pathogens among geographically clustered chronic disease patients or linking chronic disease incidence with seasonal activity among pathogens or their vectors can provide useful support for hypotheses of infectious involvement in chronic disease. Learning about the causes for these and other chronic diseases is an exciting challenge with a significant role for an applied understanding of evolutionary biology.

Understanding Variation in Populations

Individuals do not evolve, populations do. When someone is seriously ill, they want and need immediate, individual treatment. However, all the genetic traits of primary concern to public health officials and physicians evolve among populations, and discovery of the causes and consequences of that population's genetic variation is key in understanding the origins and possible management of disease.

Population geneticists have the complex task of identifying relevant genes, quantifying their variation in populations, and using those data to study evolutionary processes and their consequences. Put simply, the main processes driving change in the genetics of populations for all organisms are (1) mutation in DNA and RNA, (2) natural selection via differential reproductive success of individuals owing, all or in part, to their genetic differences, and (3) genetic drift, which is the random change in frequency of alternative genetic traits.

Virulence

Virulence is defined as the degree of sickness caused by a pathogen. Both increases and decreases in virulence for pathogen populations can be understood as part of an evolutionary response enhancing the reproductive success of pathogens. In many cases, virulence is a by-product of high rates of pathogen population growth within infected hosts. Population size can be measured by assessing the viral concentration, or titer, in blood samples. High concentration of circulating viruses can result in the destruction of many host cells, and a massive deployment of the host's immune-system defenses. These defenses can include painful inflammation of tissues, fever, and the manufacture and use of a series of remarkable molecular weapons ranging from interferon (a protein that interferes with virus reproduction), protein antibodies able to recognize and bind to specific pathogens, and specialized white blood cells capable of finding and killing pathogens by ingesting them or dissolving their cell walls. Our bodies routinely dispatch most pathogens with ease; however, unfamiliar pathogens can wreck havoc as they proliferate, despite defensive measures.

High viral titers indicate rapid reproduction and an evolutionary bet that virus progeny will be able to disperse and infect new hosts before their current host dies. This view supposes that virulence tends to decrease as dispersal opportunities decrease and increase as dispersal opportunities increase.[23] This follows from an understanding that natural selection operates on individual variation within populations in reproductive rate and dispersal mechanisms. For example, pathogens transmitted by mosquitoes, such as yellow fever virus, or by water, such as *Vibrio cholerae,* the cholera bacterium, can afford (in an evolutionary sense) to be virulent, confining their hosts to bed, because transmission does not require hosts to be mobile. Other pathogens, however, require close physical contact among old and new hosts for transmission. Examples of these include rhinoviruses, which cause the common cold, and herpes viruses, which cause cold sores. If these pathogens are so virulent that they incapacitate their hosts, they reduce the chances for their progeny to disperse directly to new hosts. Where human population densities are high, pathogens are also less reliant on ambulatory hosts for dispersal, and virulence may increase. A general correspondence between the ease of transmission and virulence for many pathogens, as noted above, indicates their influence on each other over evolutionary time. Variation in traits affecting transmission and virulence within populations provides the raw material for natural selection.

This evolutionary view of virulence as a natural response to transmission opportunities, though simplistic, has obvious practical implications. If we can reduce transmission opportunities among host populations, we might also reduce virulence and abundance of their pathogens. Reducing mosquito populations near human habitations by reducing their favored breeding sites and reducing human exposure to mosquitoes are important in managing mosquito-borne pathogens such as yellow fever virus, dengue fever virus, and West Nile virus. Similarly, treating sewage hygienically and keeping water supplies free of pathogens are important in managing abundance and virulence of diarrhea-causing pathogens. Diarrheal disease remains a leading cause of death around the world. Pathogens involved include bacteria in the genera *Vibrio, Cryptosporidium, Campylobacter, Salmonella,* and *Shigella,* as well as the protozoan *Giardia lamblia.* Although intuitively obvious, these public health practices are dictated by the principles of evolutionary population genetics. Reduction in reproductive and transmission opportunities for pathogens can decrease their population size and potentially induce natural selection to favor decreased virulence.

Left to its own devices, natural selection may favor increased virulence where transmission opportunities increase. The 1918 influenza A pandemic appears to have started among U.S. army recruits and to have increased dramatically in virulence as it spread among soldiers crowded together in battle trenches, barracks, and hospitals along the Western Front in World War I, ultimately killing more than 21 million people worldwide.[24] Subsequent influenza A strains with the same H1N1 combination of surface proteins (hemagglutinin type 1 and neuraminidase type 1) have been less virulent. Though

HIV-1 and HIV-2 represent two independent host-species shifts into humans of two different primate immunodeficiency viruses, HIV-1 has been by far the more virulent. Not at all coincidentally, HIV-1 is much more closely associated with population centers experiencing increasing levels of sexual contact among individuals, providing increased opportunity for transmission.

The views outlined above reduce the evolution of virulence to tradeoffs between the frequency of transmission and virulence, and this is overly simplified.[25] There are many other factors that can influence evolutionary change in virulence. For example, virulence may increase more quickly in populations without resistance compared to those where at least some individuals are resistant. Virulence may also increase relatively quickly where the disease-inducing parasite reproduces sexually or where host species reproduce asexually, because sexual reproduction is generally better for removal of damaging mutations. Both transmission and virulence are complex phenomena whose relationship to each other and to variation in life history traits of pathogens and hosts remain poorly known. Whether or not we can tame virulent pathogens by changing human behavior to reduce pathogen transmission is not known, and the task seems daunting at best, given the difficulty of legislating human behaviors.

The notion of simple tradeoffs between transmission opportunity, reproductive rate, and virulence or resistance evolution will undoubtedly be revised as more is learned about microbial lifestyles and social behavior. For example, many bacteria are known to be able to sense local population density, an ability called "quorum sensing." This is accomplished by chemical signaling. Some use this information to delay expression of virulence genes until their population size is sufficient to withstand the host's immune response.[26] Understanding the mechanism of quorum sensing and blocking it may prove useful as a means for preempting virulence increase in some bacteria.

Resistance

Resistance to drugs is favored by natural selection acting on the variation within pathogen populations. Resistance can develop quickly as a consequence of the high variability and rapid reproduction of many pathogens. Penicillin was first used against human bacterial infections in 1943 and was highly successful. However, strains of resistant bacteria had arisen only three years later. HIV populations in a single person can evolve resistance to antiviral medication within a matter of months or even weeks because of the virus's high reproductive rate and intensive natural selection.

Drug resistance in pathogens may evolve when chance mutations conferring resistance are passed on to progeny. Resistance may also arise when genes or gene fragments conferring resistance are traded (passed horizontally) among individual bacteria. In the presence of the drugs targeted against pathogens, individuals with resistance will be favored while others in the population without resistance will fail to reproduce. Understanding how drug resistance evolves and how it might be stalled has practical implications for everyone.

Prophylactic use of antibiotics to guard against possible infection is widespread. It has been shown to reduce the likelihood of infection following bone marrow transplants or cardiovascular and orthopedic surgery and to prevent ear infections in children. It can also protect children ill with a disease such as measles, which weakens the immune system, increasing susceptibility to other infections. Yet prophylactic application of antibiotics promotes the evolution of resistant strains, which may cause even more severe problems for both individuals and populations. Though the practice is justifiable at times, there are increasing calls to avoid it except where carefully controlled and demonstrated to be effective.

Prophylactic use of antibiotics is also widespread and problematic as practiced by farmers, who put antibiotics in feed to enhance the growth rates, size, and health of their livestock. This will tend to select for strains resistant to the particular antibiotics used, which may then infect humans or confer resistance to particular antibiotics by contributing resistance genes via horizontal transfer to human bacteria. Despite bans on prophylactic antibiotic use in livestock feed, their use is increasing as many pig and cattle growers, in particular, are able to subvert the regulations. For a near worst-case scenario, imagine a human taking antibiotics for a stomach infection who eats beef still carrying antibiotic-resistant bacteria. Upon infecting the human gut, the resistant bacteria find a system largely devoid of competitors or predators, accelerating their reproductive success and spread. Bacteria that develop drug resistance in livestock on antibiotics, such as VRE (Vancomycin-resistant *Enterococci*), are known to be able to pass from animals to humans by direct contact or through contaminated meat. They then are able to transfer their antibiotic resistance via plasmid vectors to similar bacteria in the human gut. This is of particular concern because Vancomycin has been one of the most effective antibiotics until recently.[27] Initial benefits for human health and for livestock production from prophylactic antibiotics have lured people into an escalating need for additional antibiotics as pathogens evolve resistance to existing ones.

An evolutionary view underscores the need for the wise deployment of antibiotics. Wise use includes careful adherence to antibiotic prescriptions. If antibiotics are taken only long enough to weaken or kill some but not all pathogenic bacteria from a system, that also selects for resistant individuals that can spread. This occurs when patients who are prescribed antibiotics see temporary improvement in their health and lose interest in completing the lengthy regimen of antibiotics required to drive the responsible bacterial population to extinction. Wise use means consideration of the risks involved in unwittingly selecting for resistant pathogens, not just in people but in domestic and wild animals as well.

An understanding of the evolution of resistance contributed to the rationale for development of combination drug therapy. For example, AIDS patients taking three antiviral drugs are more likely to forestall the evolution of viral resistance than are those taking a single antiviral drug. One mutation providing resistance to a single drug can arise and be selected for much more readily than can multiple, different mutations simultaneously providing resistance to different antiviral medications. At present, combined drug therapies for HIV infection are the most successful treatment known.

Genetic and Environmental Contributions to Disease

Diseases that are not a result of infectious pathogens can stem from genetic mutations, environmental factors such as exposure to toxins or unhealthy diets, and combinations of the two. Measuring genetic contributions to disease is complicated, because single genes may influence multiple traits (pleiotropy) and single traits may be influenced by multiple genes (polygeny). Environmental contributions are similarly complicated, and the interaction of environmental and genetic factors add yet another dimension to the challenge of understanding noninfectious disease origins. Examples of multi-factorial diseases include diabetes, coronary heart disease, breast cancer, Alzheimer's disease, and schizophrenia. Biologists use the methods of quantitative genetics to assess the relative importance of genetic versus environmental factors in determining the traits of organisms and to understand how the variation in traits among populations translates into evolutionary change. Many risk factors for disease, whether genetic or environmental, have only small effects individually. This underlies the view that many genes, each with small effects, influence the variation of disease susceptibility within populations.

For example, quantitative genetics of schizophrenia, focusing on disease variation in populations and closely related family groups, indicates that risk for schizophrenia is 10 percent for siblings of victims compared to 1 percent for the general population; 45 percent of identical (monozygotic) twins both have the disease, compared to 5 percent of nonidentical twins. Heritability, which is the fraction of population variation in traits that are a result of genetic effects, is estimated at about 80 percent for schizophrenia. The specific genes involved are not yet known. Environmental influences are not known either, though some evidence suggests that problems in early brain development, possibly due to malnutrition or viral infection during the mother's pregnancy, may contribute. Severe stress in later life has also been implicated.

The key aspect of evolutionary thinking in quantitative genetics is the focus on population variation and degree of heritability for traits. Like all other traits of organisms, those involved in the genetics of disease susceptibility are subject to the evolutionary forces of natural selection and genetic drift operating within populations. Thus, understanding the quantitative genetics of disease is largely the same as studying the evolution of complex traits. Evolutionary biology provides the necessary context for interpreting observed patterns of genotypic and phenotypic variation in populations.

Understanding cardiovascular disease susceptibility is a bit further along than understanding schizophrenia. Over sixty candidate genes for involvement with lipid metabolism, carbohydrate metabolism, blood pressure regulation, and homeostasis have already been identified. This does not even include genes involved with heart muscle form and function. If specific alleles for different genes can be linked to risk, this provides a basis for diagnosis and possible treatment. Quantitative genetics as well as phylogenetic analyses of just a few of these candidate genes, such as LPL (lipoprotein lipase), sampled across human populations already indicate that this will not be easy. Many alternative forms (alleles) for relevant genes are present in populations, and there is a complex relationship between genetic and phenotypic variation.[28]

Applications of Artificial Selection

When the topic of artificial selection comes up, most people think of domesticated animals and crops; however, the utility of artificial selection is much broader. Artificial selection and its operative evolutionary principles have applications for the fields of health and medicine as well. As with domestication of species for agriculture, applications in health and medicine have progressed from an intuitive, experimental approach focusing on whatever works to a highly targeted approach selecting on and enhancing particular molecular genetic traits of individuals.

Vaccine Development

The application of evolution via artificial selection is showcased in the intentional evolution of attenuated (weakened) viruses for use as vaccines against disease. Attenuated viruses are those strong enough to elicit antibody production and lasting immunity, but weak enough to avoid causing disease. The earliest record of intentional infection with mild forms of smallpox virus are from China around 1200. Pus or scabs were taken from sores in someone mildly ill and rubbed into small cuts of someone not yet infected, or else a ground powder of the infectious agents was blown up the nose of the uninfected person. This practice likely followed from the observation that individuals accidentally exposed to pus of smallpox victims, if they survived, appeared immune upon exposure at a later date.

In eighteenth-century Europe a modification was practiced in which cowpox virus, an animal pathogen, was passed serially among humans by taking pus from cowpox sores on one person and using it to infect another, whose sores were used, in turn, for pus to infect another person. Again, the purpose was to provide immunity to the much more virulent smallpox. A benefit of using cowpox was its greater frequency and potential availability in communities prior to a human smallpox outbreak. In 1796, Edward Jenner publicly demonstrated this use of cowpox to provide immunity to the closely related human pathogen, smallpox. Jenner's demonstration predates any direct knowledge of viruses or evolutionary relationships among them; however, the practice worked precisely because of the close evolutionary relationship among poxviruses from cows and humans. This lesson in applying knowledge of evolutionary relatedness has been routinely applied. Similarities among poxviruses (a diverse assemblage infecting many vertebrates) allows antibodies elicited by some to recognize and protect against disease caused by others. Cowpox viruses intentionally introduced and then passed from human to human are strong enough to elicit production of antibodies for poxviruses, but generally weak enough to avoid fatal pathogenesis.

Explicit artificial selection for attenuation was also used in developing the BCG (bacille Calmette-Guérin) vaccine against tuberculosis. In the 1920s, Albert Calmette and Camille Guérin found that *Mycobacterium bovis* selected for growth in culture on potatoes eventually lost its virulence for animals but retained its ability to stimulate protective immunity against

Table 3.1 Attenuated vaccines made from wild bacteria and viruses by mimicking natural selection in the laboratory

Bacterial vaccines	Viral vaccines
Anthrax[a]	Adenovirus types 4 and 7
Chicken cholera[a]	Hepatitis A[b]
Cholera[b]	Influenza A and B[b]
Salmonella[b]	Measles
Shigella[b]	Mumps
Tuberculosis (BCG)	Parainfluenza type 3[b]
Typhoid[b]	Poliomyelitis types 1, 2, and 3
	Rabies[b]
	Rotavirus[b]
	Rubella
	Smallpox
	Tick-borne encephalitis[b]
	Varicella
	Yellow fever

a. No longer used.
b. Still under investigation.

tuberculosis in people. BCG is generally given within 15 days after birth to children where tuberculosis is endemic.

An attenuated, live viral vaccine against yellow fever in humans was developed via artificial selection in 1935 by taking the French neurotropic strain (FNS) of the yellow fever virus and passing it, serially, through mouse brains. Another yellow fever virus strain (17D) was attenuated by serial passage through rhesus monkeys and mosquitoes, followed by passage through both mouse embryonic tissue cultures and chick embryos. Because the FNS strain retained greater virulence, the 17D strain, further artificially selected on to improve its stability at high temperatures, is currently used in making the vaccine. This vaccine is credited with saving many lives over the past sixty years, though wild virus populations continue to thrive in parts of Africa and South America.

The same artificial selection approach is used in current attempts to make a vaccine against human rotavirus infection from cow and rhesus monkey rotaviruses. Rotaviruses cause nearly a million deaths a year, primarily in infants, due to dehydration from diarrhea and vomiting. Similarly, a vaccine against human parainfluenza, a respiratory infection, is being developed from a closely related paramyxovirus of cows. Other examples are given in the table.

This is not to suggest that vaccine development is simple or that evolution of wild or artificially selected virus strains stops. Evolutionary change can bring renewed virulence as viruses evolve novel antigenic properties and evade or overpower the host's previously effective immune system. Attenuated viruses can and sometimes do revert to virulence. This has been seen in some polio viruses, where evolutionary change has restored virulence in previously attenuated forms. An evolutionary view makes obvious the ideal of attempting to inoculate whole communities simultaneously, or nearly so. If virulence reevolves, most will have already developed some immunity based on the attenuated innoculate. Other problems arise when vaccine strains are insufficiently weakened in their preparation, inducing serious illness in recipients, or if contaminated with other pathogens, potentially from cells used to culture vaccines or from tissues used in isolating vaccine candidate strains.

Some of these problems are circumvented by genetically engineered vaccines, an extreme form of artificial selection. Chimeric viruses or bacteria can be constructed in the lab, in which a gene from a virulent pathogen is selected and added to a nonpathogenic virus or bacterium. Following inoculation, the inserted gene will be expressed, inducing antibody production and, hopefully, subsequent immunity. Reversion to virulence is not an issue because the chimeric entity does not carry the full virulent pathogen genome, and contamination with other source pathogens is obviated because the construct is synthetic and does not involve isolates from blood or other tissues.

Identification of the specific mutations responsible for the evolution of attenuation or reversion to virulence is possible using comparative analyses. This allows monitoring of the stability of the relevant mutations during vaccine manufacture and field trials. Also, specific attenuating mutations can be inserted to engineer improved candidate vaccine pathogens, although the process remains difficult.[29]

Vaccines developed via artificial selection or genetic engineering, including those for diphtheria, tetanus, polio, *Haemophilus influenzae,* hepatitis B, chicken pox, measles, mumps, rubella, pertussis, and pneumococcus, have saved millions of people. Even so, the targeted pathogens have not been driven to extinction (with the apparent exception of smallpox), and they continue to ravage human lives. There are many reasons for this, including their presence in nonhuman hosts and the difficulty of coordinating and funding worldwide eradication programs for humans, but one of particular relevance here is pathogen success in adapting to changing environments. Their ability to evolve rapidly is their hallmark. Pathogens do not wait passively as we attempt to deliver a knockout blow. To be successful in managing or preempting infectious diseases, we will have to better understand their past history and current mechanisms of evolutionary change. We must also recognize that altered or novel pathogens will continue to arise, and the more we know about their phylogenetic origins, their ecology, and their molecular evolution, the better informed and more successful our management efforts will be.

Directed Evolution of Molecules

The same general approach of artificial selection applied so productively to crops, livestock, and pathogens (as vaccines), can be applied to molecules of RNA, DNA, and protein. The artificial selection of molecules is generally known as "directed evolution," and may be described as a three-step process. First, variant molecules are created either directly by synthesis or indirectly by intentional mutagenesis of molecules of known (or suspected) function. Second, the variant molecules are assessed for the desired performance attributes. Third, top-performing molecules are selected and passed on to the next bout of mutation and performance assessment. Following each bout of artificial selection, molecules performing at a lower level are eliminated from the population. As with artificial selection on plants and animals, humans determine which individuals are reproduced.

There are many variations on how to perform these steps, but the evolutionary approach is the same. So far, most directed evolution experiments have been performed with RNA or DNA molecules, because they can be replicated directly, whereas proteins require RNA and DNA intermediates and more complex molecular interactions. Many or most biological tasks targeted by directed evolution experiments are performed by proteins. Thus, working with proteins remains a long-term goal.

Often the initial choice of the molecule or gene to be subjected to directed evolution is based on a determination of its evolutionary relatedness to the same (homologous) gene in another species in which the function of that gene has been experimentally determined. The working premise is that homologous genes in different species will have identical or similar functions as a result of their common ancestry. Variant copies of the genes of interest can be generated by replication with random mutations or mutations inserted intentionally at specific sites thought to be functionally important. Alternatively, gene regions can be copied and intentionally shuffled. As more organisms and genomes are sequenced and added to databases, there are more known homologous genes from different organisms. These homologous genes, with similar capabilities, can be recombined or shuffled and then artificially selected for enhanced function under different environmental conditions.

Screening variant genes and their products for performance involves a diverse set of chemical assays and physical screens. The variable success of different approaches in generating mutations and screening the resulting products reflects our lack of knowledge regarding the links between DNA sequence, on the one hand, and molecular structure and function, on the other. Fortunately, such knowledge is not a prerequisite for directed evolution. An evolutionary approach allowing random mutagenesis of target molecules may find improvements based on changes at sites and in regions unsuspected of being important previously. This approach mimics how natural selection operates on natural variation in the wild. Further, directed evolution experiments may find multiple variants with improved function and find them more efficiently than targeted mutagenesis potentially based on misconceptions linking sequence and function. The method is outlined in the accompanying figure.

Not surprisingly, directed evolution of specific proteins used in treating human diseases has sparked commercial interest. The goal is to modify or create therapeutic proteins with greater efficacy, fewer side effects, and less cost compared to available alternatives. The classes of therapeutic proteins include: *antibodies,* which bind to pathogens, identifying them as foreign entities (antigens) to be eliminated by the immune system; *cytokines,* which regulate the immune system and its response to inflammation; *hormones,* which are critical to normal development, growth, and metabolism; and *enzymes,* which facilitate many processes in the normal functioning of cells. As one example, an antibody in current development using the directed evolution approach, AME-527, binds and neutralizes human TNF alpha, thought to contribute to various inflammatory diseases, including rheumatoid arthritis. The many challenges in directed evolution include overcoming weak binding to target antigens, low potency, unfavorable side effects, instability, and excessive production costs. Potentially, all these factors can be improved incrementally via the artificial selection process embedded in directed molecular evolution.

Directed evolution has been used in the development and optimization of single-stranded DNA or RNA molecules called "aptamers," which can bind to pre-defined, specific target molecules. The folded, three-dimensional shapes of aptamers allow tight binding in a lock and key configuration analogous to an antibody-antigen interaction. Aptamers have

Figure 3.7 Illustration of directed evolution applied to development of enzymes and drug compounds. Change in the enzyme's catalytic fitness (ability) is shown on the vertical axis and change in its DNA sequence is shown on the horizontal axis. (i) Incremental improvement by repeated error-prone DNA sequence copying or DNA shuffling (mixing DNA pieces from related genes taken from different species) can find the local catalytic fitness maximum. (ii) Targeted mutation of relevant enzyme regions, such as the active site, may find new peaks or functions in fitness by identifying multiple mutations that work well together.

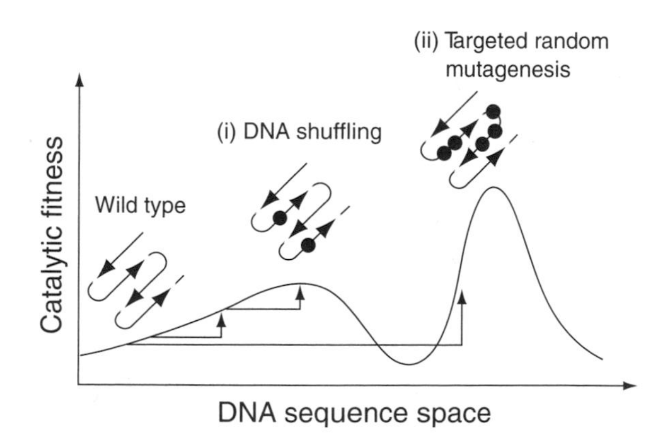

several advantages compared to antibody-based affinity, including greater ease in production and greater stability due to the chemical properties of nucleic acids versus amino acids. Use of aptamers, primarily to inhibit target proteins, is still in its early stages, though there is hope that they will be used eventually to treat many diseases, including AIDS and cancers. Aptamers have been produced for nearly all classes of known molecules, with binding abilities optimized through forced mutation and artificial selection. Delivery of aptamers to specific, target cells remains a primary challenge.

Biochemists have used an approach analogous to genetic recombination called "combinatorial synthesis" in drug discovery. By synthesizing and mixing chemical compounds, combinatorial synthesis yields a huge library of candidate molecules to perform a particular function and provides raw material for artificial selection in a parallel fashion to recombination and natural selection in organisms. One example is recent work to find a replacement for Vancomycin, a once highly effective bacterial antibiotic. The Vancomycin molecule was originally derived from the bacterium *Streptomyces orientalis,* found in soils from India. Vancomycin evolved as a defense against other bacteria by inhibiting formation of the bacteria's cell walls (as does penicillin). However, as with other antibiotics, resistance is increasingly common. K. C. Nicolaou and colleagues demonstrated the feasibility of a form of combinatorial synthesis in development of effective Vancomycin analog molecules working in tethered pairs.[30] They synthesized the chemical building-blocks for Vancomycin individually and incubated them together with molecules able to link them. The products of the linked Vancomycin analogs were then screened and artificially selected based on their performance as antibiotics. The results indicate promise, though problems remain.

Directed evolution is also being used to enhance the utility and environmental safety of enzymes in industry. Proteases and cellulases, optimized via directed evolution, have been used to replace pollutant phosphates used in laundry detergents. Evolved amylases and pectinases have replaced sodium hydroxide as a detergent and softener in some textile factories, and evolved lipases have replaced some chemical emulsifiers in bread making.[31] In the long term, development of environmentally friendly and more readily digested enzyme replacements could have a positive impact on human and environmental health.

Artificial selection on molecules to enhance their performance promises to be a fruitful enterprise, with the potential to alter human lives in ways that might rival the impact of artificial selection on plants and animals over the past 10,000 years. Recognition of this potential underscores the real-world value of applying what we have observed and learned about the evolutionary process.

Health and Prior Human Evolution

In preceding sections we applied the concepts of evolutionary biology to help us understand and respond to disease. Let's take a different perspective here, and consider how human evolutionary history itself affects our health. This view will not necessarily reduce illness, but it can help explain our susceptibility to health problems.[32]

Bipedalism, or walking upright, distinguishes humans from other primates, and based on fossil evidence, including 3.5 million-year-old footprints from northern Tanzania, our ancestors have been bipedal for about 4 million years.[33] Bipedalism provides certain benefits, including freeing the arms and hands for carrying and manipulating objects, but it has its costs in placing significant novel stress on bones and joints not adapted for this task during the previous 240 million years of mammalian quadruped evolution. Frequent lower back, hip, and knee malfunction and pain are influenced by bipedalism, because the weight previously distributed across four limbs has been distributed to two.

Along with bipedalism, the human pelvis evolved to be proportionally shorter and broader, providing support for the torso and internal organs. As part of this morphological change, the birth canal narrowed, constraining the size of infants' heads at birth. These changes took place despite a concomitant increase in brain and skull size. The increase in the size of the human brain was facilitated by accelerated growth rates, relative to other primates, during the first year after birth. This still leaves the average skull size of human infants at birth only slightly smaller than the birth canal, resulting in a relatively high incidence of complications during human births. By comparison, the birth canals of nonhuman primates allow much easier passage of infants.[34] An evolutionary view in this case does not lead to recommendations, though it does provide a cogent explanation and some understanding for the existence of various problems in childbirth.

Other changes in human reproductive biology are apparent in comparing traits in modern hunter-gatherers, indicators for the reproductive traits of women in pre-agricultural times, to the same traits in modern women from affluent societies. Onset of menstruation is later in hunter-gatherers (16 compared to 12.5 years), though their first births are earlier (at 19 compared to 24.5 years of age). Average number of births is greater in hunter-gatherers (6 compared to 1.8), time spent nursing is greater (2.5 to about 0.5 years), and onset of menopause is also earlier (at 46 compared to 52 or so years of age). These changes in reproduction parallel increased risk in more affluent societies for ovarian, uterine, and breast cancers.[35] The cause remains to be proven and environmental effects may outweigh genetic factors. However, increase in the exposure of reproductive tissues to estrogenic hormones and the resultant increase in the total number of lifetime ovulations and cycles of breast cell growth in preparation for lactation may stimulate greater incidence of cancer, because more rounds of cell division and DNA replication provide more opportunity for cancer-causing mutations to arise. Earlier onset of menstruation, fewer pregnancies, and less time spent nursing in some societies all result in greater exposure to estrogenic hormones and more cell divisions for reproductive organs.

Similarly, prostate cancer in males appears correlated, by some accounts, with lifetime levels of androgens, including testosterone. As an extreme example, castrated males are not known to get prostate cancer. Prostate cancer cells require testosterone, especially in their early growth stages, which is why anti-androgen therapy has been a primary approach to treatment.[36] An evolutionary approach to this problem might include attempts to reduce levels of hormone exposure and frequency of cell proliferation associated with menstruation and lactation in females and sperm production in males, approximating those experienced during pre-agricultural times. Such a hypothesis remains speculative, however, and at best provides a partial explanation. Genetic and environmental factors are certainly involved and integral to a more comprehensive understanding.

The changes in average human life span is another trait influencing human health. Average life spans are approaching 80 years in developed countries and 65 years worldwide. This may be compared to average life spans around the world of about 47 years in 1900 and just 24 years in 1800. This last estimate is quite similar to the estimate of about 20 years for Neanderthal humans living 60,000 years ago, suggesting that much of the increase in average human life span has arisen in the past 200 years. The increase is largely due to improvements during the past 150 years in hygiene and protection from infectious disease sources, as well as lower infant mortality.

Increasing longevity is correlated with increasing incidence of diseases such as heart disease, stroke, high blood pressure, diabetes, Alzheimer's disease, and arthritis. Though not restricted to older humans, these diseases are much more common among them. Why are they most prevalent among older individuals? An initial observation notes that, in earlier times, most people did not live long enough to develop these diseases. The observation contains a clue not missed by evolutionary biologists. Selection promoting reproductive success of individuals is strongest for traits expressed before and during the peak reproductive years, and less so, though not absent, for traits that do not appear until after the reproductive years.

The question about diseases of old age can be expanded. We might ask: Why do organisms age and die? The only scientific explanations for the existence of aging (or senescence) are evolutionary, and these center on tradeoffs between natural selection acting directly and strongly on genes that influence reproductive fitness of individuals up to and during their most reproductive years, but acting only weakly if at all on traits affecting individual health in later life stages. One form of this argument, known as antagonistic pleiotropy, supposes that genetic traits conferring reproductive advantage may negatively affect health in later years.[37] Alternatively, traits with negative effects in later years may be selectively neutral during early stages of life, allowing them to persist over evolutionary time. Life span and the onset of senescence can change and evolve just as other life history variables do. Artificial selection experiments on fruit flies have demonstrated that longer life spans can be selected for and can yield populations with heritable, increased life spans. However, to the extent that senescence and diseases of old age stem from genetic traits that have different effects at different stages of life, we can expect there will be limits to human life span increase.

An evolutionary view of public health and medicine strives for realism, considering organisms in their environments as the unfinished result of millions of years of evolutionary mistakes, innovation, and improvement. The more we understand about the origins, capabilities, and limitations of pathogens, the better prepared we are to minimize their negative effects on us. The better we understand our own capabilities and historical constraints, the better prepared we are to manage our own health. By mimicking natural selection via artificial selection on some bacteria, viruses, and individual molecules with useful properties, we can enhance human health. The possibilities are only beginning to be explored. Yet an evolutionary view also shows that the health and lives of individuals are necessarily ephemeral. In the grand sweep of the evolutionary history of life, change is continual. An individual and its particular genetic configuration is static, however; only populations can evolve.

4

EVOLUTION AND CONSERVATION

Life on Earth is about 3.8 billion years old. In that span of time, life forms have produced a flood of variants, with billions of individuals of millions of kinds testing their abilities and luck in survival. These variants form a stream through time with succeeding generations linked by the currents of heritability. Just as a river might divide into countless rivulets as it flows into a vast, uneven delta, families of variant organisms differentiate from each other over time and develop alternative strategies for life and reproduction. The alternative paths for organismal differentiation are infinite, constrained only by the physical limitations of their bodies and genomes and, of course, the nature of their environments. Over tens of millions of years, as groups of organisms differentiate, and one lineage becomes many, novel means for survival are pioneered and the lifestyles of ancestors are left behind, potentially in the safekeeping of more conventional relatives.

It is difficult to comprehend the richness of biological diversity. Doing so requires a capacity for distinguishing the kinds and differing traits of organisms that quickly overwhelms even the most encyclopedic memory. We might as well expect an astronomer to have personal familiarity with the particulars of all the stars in the sky. Indeed, the full measure of life, past and present, remains undescribed and uncatalogued, so that even those with the desire are left to extrapolate based on the shifting ground of current knowledge.

This chapter, however, considers the conservation of biodiversity in light of evolutionary understanding, so I must at least attempt an outline of biodiversity. To begin, let's consider the diversity of surviving forms, and roughly compare just a few of their attributes to ours. Marvels have been wrought as groups of organisms have evolved over time, and the sampling of species discussed here hints at the diversity to be found in other lineages, well out of the limelight, that have evolved capabilities far surpassing our own.

What Is Biodiversity?

All life forms need energy to survive. We eat rice, potatoes, fish, and cows among other plant and animal products, and convert their chemical energy into metabolic and mechanical energy for ourselves. Eating other organisms, a lifestyle known as heterotrophy, is considered by us to be conventional, although the means to that end are variable. Andean condors may travel 200 surface miles or more in a day, soaring high above the ground on rising air masses, in search of the rotting flesh of recently dead animals. Deep-sea sponges remain in one spot, beating their flagella to create small currents bearing their bacterial meals to a filtered orifice. Leaf-cutter ants grow their own edible fungi in elaborate subterranean nest-gardens that may be 1,000 square feet in size and harbor a million or more individuals.

Using an entirely different approach, plants and algae take in sunlight, carbon dioxide, and water and, with the assistance of the light-trapping pigment chlorophyll, convert them into oxygen and sugars for their food. Practitioners range from giant kelp, some over a hundred feet tall, arrayed in undulating forests that filter their sunlight through cold marine waters, to tropical epiphytic orchids suspended from the lofty canopy and thriving without soil. Tiny cyanobacteria are capable of photosynthesis via chlorophyll as well. They form delicate chains or filaments of photosynthetic cells, lending their powers of sugar creation to a variety of hosts, including cycads, sponges, sea-squirts, and diatoms. Who would not envy turning green with chlorophyll for such a sweet relationship?

Less familiar, and all the more impressive, is the ability of many organisms to derive their requisite energy from inorganic substrates. This is accomplished by taking electrons from inorganic molecules such as hydrogen, methane, carbon monoxide, hydrogen sulfide, ammonia, nitrogen dioxide, nitrous oxide, iron(II) and manganese(II) and using them to synthesize ATP, an energy-storing compound. Creatures living in this fashion are known as lithotrophs, meaning "rock eaters," and they are ubiquitous, occurring in seawater, freshwater, sewage treatment ponds, the atmosphere, volcanoes, and deep sea hot springs. Nitrifying bacteria, such as the *Nitrosomonas* and *Nitrobacter* species found in stream beds, are examples of lithotrophs. They are willing reducers of the ammonia that is typically released by degrading organic materials. When areas of volcanic activity in the deep sea were first explored in the 1970s, scientists were amazed to find flourishing communities of crabs, clams, and marine worms surviving in pitch darkness at 3,700 meters depth. Part of the key to survival there turned out to be the lithotrophic bacteria living by reducing the sulfur spewing from hydrothermal vents. Once bacterial lithotrophs are established as primary producers, colonization by other organisms can follow. Some feed on the lithotrophs, and others such as the tube worm *Riftia pachytila* play host to lithotrophic bacteria, providing a specialized growth chamber as well as blood-borne oxygen in exchange for a portion of the bacterial ATP production.

How we get our energy is only one measure of diversity on our planet. Another that varies widely among organisms is ambient temperature. Our lineage began diverging from other primates in subtropical African savannas about 7 million years ago, and we have long been accustomed to living within the temperature range of 20°C to 30°C. If our core body temperature rises above 42°C for just a few minutes, we will likely die; if our body temperature drops to 35°C or lower, we develop hypothermia and often die. But others have evolved to have different tolerances. The archaeal bacterium, *Pyrococcus fumarii,* found in shallow marine thermal vents, grows best at body temperatures in the vicinity of 113°C (235°F), well above the boiling point of water at 100°C.[1] Not only do such hyperthermophilic bacteria manage to stave off death, they have evolved enzymes uniquely well suited for replicating DNA, a rather exacting task, at these extreme temperatures. Enzymes of any sort from most other organisms simply decompose at such temperatures. At the other end of the temperature spectrum for optimal growth in bacteria are the cold-loving, or psychrophilic, forms inhabiting alpine glaciers, polar seas, and deep-sea sediments. These organisms can grow at body temperatures as low as −10°C. Lake Vostok, nearly four kilometers below a glacier in Antarctica, contains bacteria representing at least five different genera.[2] Needless to say, we could not succeed there on our own.

Oxygen comprises 90 percent of the planet's water, 66 percent of the human body, and 20 percent of the volume of air. Oxygen is our prerequisite. For us, it is synonymous with life itself. Human brain tissue begins to die following just six minutes of oxygen deficiency (a good reason to administer CPR to someone immediately after a heart attack), and people normally inhale fifteen to twenty times per minute. However, many other life forms can live for long periods without breathing. Some marine mammals, including sperm whales and elephant seals, can stay underwater without breathing for as long as two hours. This is accomplished not by increasing lung capacity, but by increasing the efficiency of lungs in transferring oxygen to and carbon dioxide from the blood. Anaerobic fungi and bacteria thrive in the absence of oxygen, allowing them to live in seemingly hostile environments, such as the intestines of animals and the densely packed mud of the deep ocean floor.

Radioactivity exposure damages DNA and the functioning of cells. Radioactivity is measured in units called "rads," and 500 rads exposure is lethal for humans. The bacterium *Deinococcus radiodurans,* however, can withstand 1.5 million rads, or 3,000 times the amount lethal in humans, without ill effect. This organism is also resistant to high levels of ultraviolet radiation and to desiccation. The resistance to levels of radiation that the vast majority of life would find fatal are attributed to a suite of mechanisms promoting rapid repair of damaged DNA and chromosomes as well as a multi-layered cell wall rich in radiation-blocking peptidoglycans, a lattice of amino acids and sugars. Impressive as this radioactivity resistance is, it has apparently evolved as a by-product of selection for resistance to desiccation.

Another measure of biological diversity is reproduction, one of life's defining operations and a prerequisite for evolution and adaptive change. For our species and a relatively small number of others, reproduction is sexual, meaning the genes of parents are combined in the offspring. The alternative is asexual or clonal reproduction, in which offspring receive all their genetic material from a single individual. Asexual reproduction is more common, being optional or required among viruses, bacteria, and many invertebrate animals such as amoebas, sponges, and sea anemones, as well as many plants. And, if either approach can work, why not have it both ways? Many organisms, such as ferns and jellyfish, do just that, alternating between sexual and asexual phases in their life cycles. Some organisms do not bother to alternate, being hermaphrodites with both male and female reproductive structures present simultaneously. Most flowering plants are hermaphrodites, as are many invertebrate animals, such as earthworms. Some but not all hermaphrodites can fertilize themselves, a seemingly convenient circumstance, though not without its inbreeding perils.

Generation time is also highly variable. Galapagos tortoises first breed around age twenty, laying twelve or so eggs annually for many, many years—they can survive 200 years or more. Retroviruses such as HIV can reproduce within minutes of their exit from a host cell, and asexually generate 10,000 offspring in less than an hour. Viruses have ingeniously managed to co-opt the molecular capabilities of their hosts, obviating the need on their part to maintain genes and structures for many of the basic tasks of DNA replication and translation into amino acids, because this is done for them by the host.

The notion that the evolution of biological diversity has been driven by abiotic environmental factors is, at best, incomplete. This is because life plays a major role in catalyzing its own evolutionary change. The activities and interactions of organisms have continual and significant influences on their own morphology, physiology, behavior, and molecular capability. Although these influences may be invisible to the casual observer, biodiversity does not simply evolve, it coevolves. Consider that the earliest life forms we know about, going back 3.8 billion years, were anaerobic bacteria surviving in the oceans and compacted mineral soils of the early earth, where free oxygen did not exist. Things changed. The world was altered by photosynthesizing descendents of bacteria and plants. As a direct result of the prodigious activity of cyanobacteria, primarily, and plants, secondarily, free-oxygen levels rose from nonexistent to about 20 percent of the biosphere today. Their lives and cumulative effect over the past 2 billion years have been transformative, literally giving breath to life and stimulating the evolution of all oxygen-dependent organisms.

The biologist Richard Lewontin has argued persuasively that organisms play a large role in their own evolution, using the beaver as one example.[3] Beavers, descended from terrestrial rodents, are poorly suited for living and traveling in terrestrial environments. They have heavy bodies and a paddle-like tail supported by short legs with webbed feet. How did this come about? When beavers disperse, they generally move into terrestrial habitats with small to midsize streams. They build dams, sometimes hundreds of feet in length, in these streams to create the ponds that provide the deeper water they use for protection from predators, like wolves and lynx, and for construction of their lodges, where they raise their kits. Thus, over millions of years, the activity and behavior of beavers has helped to shape their own morphological and genetic adaptations to an aquatic lifestyle. Further, consider how beavers influence other organisms. A young beaver seeking a new territory of its own moves upstream from that of its parents, potentially into a broad valley with several large stands of quaking aspen trees and a stream barely a foot wide meandering through the sagebrush and grasses. If successful, over a period of years the beaver creates several ponds, each twelve feet deep and sixty yards across in places, supporting populations of three or four species of fish that could not have survived there previously, a pair of osprey that feed on the fish, numerous aquatic plant species, an itinerant moose that feeds on the plants, crustaceans, algae, a multitude of aquatic breeding insects, summer populations of insectivorous bats, tree frogs, bull frogs, nesting waterfowl, and so on. All of this stemmed from the divergent behavior and lifestyle of a formerly conventional, terrestrial rodent.

Organisms play an influential role in the course of evolution, and the idea that biological diversity results simply from DNA replication errors being sorted for success by the abiotic environment misunderstands the key contribution of life to its own history. Biodiversity presents a complex history, in which the lives and abilities of descendents are based to a large degree on the activities of ancestors. This view also helps to show why many life forms need each other to survive.

Reckoning Biodiversity

Biodiversity is usually considered on three levels: genes, species, and ecosystems. This corresponds roughly to the kinds of life forms, their genetic instructions, and their interactions within communities of life forms. It must be admitted at the outset that the integrated nature of life makes the recognition of discrete units difficult. However, there is no alternative but to try, and we may expect that our abilities will improve with experience.

DNA was shown to be the material basis of inheritance in 1944 and to consist of a double helical structure with two complementary strands of nucleic acids in 1953. Those seminal discoveries opened the way for the elucidation, some fifty years later, of the complete sequence of the human genome. Knowing the sequence for the complete genome is a far cry from knowing how the genome is used to grow and sustain an individual; however, it is a large and necessary step along the way. The known sizes of organismal genomes, containing all their genetic information, range from a high of 133 million base pairs in the marbled lungfish *(Trichoplax adhaerens)* to lows of 0.50 million base pairs in the archaean bacteria *Nanoarchaeum equitans* and 0.003 million base pairs in some hepadnaviruses, a group that includes hepatitis B virus. This represents a 44,300-fold range. In many groups, especially plants and animals, large parts of the genome are repetitive DNAs or genetic elements without known function. The numbers of genes estimated range from 50,000 or so in some plants to about 30,000 in humans, 12,000 in fruit flies, 500–6,000 in bacteria, and 1–300 in viruses.

Genes were initially defined as contiguous DNA sequence strings encoding a single functional protein. But this is simplistic, because some sequences encode multiple proteins, some only a protein subunit, many require others for functionality, and some regulate but do not encode protein. As a result, genes are loosely defined as hereditary units, and descriptors used if desired. Here, I outline two different ways to organize the diversity of genes. The first way is phenetic, classifying genes based on similarity in function or in nucleic acid sequence. As a start, genes might be considered as either protein-coding or not protein-coding. Non–protein-coding genes include those coding for structural RNAs (ribosomal RNAs, transfer RNAs), messenger RNAs, small nuclear ribonucleoproteins that are involved in the splicing of eukaryotic mRNAs, repeated DNA sequences, and introns, which are the intervening DNA sequences between protein-coding DNAs. In addition, there is a diverse set of mobile genetic elements, capable of moving from one chromosomal location to another within a single genome, or, in the case of bacteria, to or from a plasmid, which is an extrachromosomal genetic unit. Protein-coding genes constitute the proteome. The so-called housekeeping genes are those protein-coding genes critical for cell structure and function, and the list of them grows rapidly as more is learned about cell function. Most protein-coding genes have tissue-specific, age-specific, or environment-specific roles, and they can be classified accordingly. This sort of classification, requiring data on gene expression patterns, is in its infancy, however.

The second approach to understanding the diversity of genes is classifying them phylogenetically, based on common ancestry. Genes are reproduced every organismal generation and therefore evolve over time in much the same way that organisms do. However, the evolutionary history of individual genes are not necessarily identical with each other, or with that of the organismal lineages or species carrying them. Within a single organismal lineage, any particular gene may evolve multiple forms (alleles) in different individuals. These variant forms may further differentiate among populations. Genes may duplicate, analogous to asexual reproduction, and they may go extinct via gene loss. Genes may gain, lose, rearrange, or swap pieces of themselves with other genes accidentally during replication. In rare instances, genes may be moved by viruses between species.

Genes may also move between genomes in individuals having more than one. To see such an organism, look in the mirror. Humans have both mitochondrial and nuclear genomes, resulting from an ancient association between an early Eukarya lineage, responsible for most of the nuclear genome, and an early Bacteria lineage, responsible for the mitochondrial genome. The ancient association worked out so well that it is now obligate; there is no going back. Many instances of genes moving from the mitochondria into the nucleus are known for eukaryotes, as are a smaller number of instances of gene movement in the opposite direction.

Now, recognizing that genes themselves can evolve via duplication, loss, and shuffling of segments independently from the organismal lineage carrying them, imagine a three-dimensional phylogenetic tree for a group of organisms in which the hollow tubular branches are filled with lines designating the evolutionary histories of the individual genes comprising a single genome. One could peel back the skin of the branches to expose the lines showing the branching pattern and diversification of individual genes.

Starting at the tips of the branches and looking back through time, alternative forms of genes would coalesce to the forms of their ancestors. You could trace the lines back in time representing genes related by duplication and watch two gene lineages become one. If, for example, you were looking at the one-thousand or so different odorant receptor genes within the human lineage (many of which have become nonfunctional in humans, but not in dogs) and traced them back in time, you could watch as pair after pair joined until you had one or a few precursors, probably having a different function, in some ancestral mammalian lineage. You could do the same for many vertebrate gene families, including the globins functioning in circulation of oxygen, histones with a role in folding and coiling DNA on chromosomes, and the *hox* gene family, which plays a crucial role in development. In some cases, entire clusters of genes are duplicated simultaneously in a single copying error. Occasionally, you would see that whole genomes have been duplicated. You could also see lateral transfer of genes, moving horizontally from one branch of the tree into another. Research over the past several decades has shown lateral transfer of suites of genes among bacteria to be an important mechanism in the evolution of antibiotic resistance, increased pathogenicity, core functions such as replication and transcription, as well as novel physiological and behavioral traits.[4] If this hindsight were perfect, you could trace the origins of all extant genes in a particular genome from a smaller set of precursors, going back to the beginnings of life.

Much of this phylogenetic history for the constituents of

Figure 4.1 Evolution of genes and genomes within organismal lineages. Lines denote evolutionary history of genes illustrating duplication of genes and whole genomes, gene loss, lateral gene transfer (between species), and exon shuffling, which is the mixing of protein-coding blocks of DNA among different genes mediated by transposons.

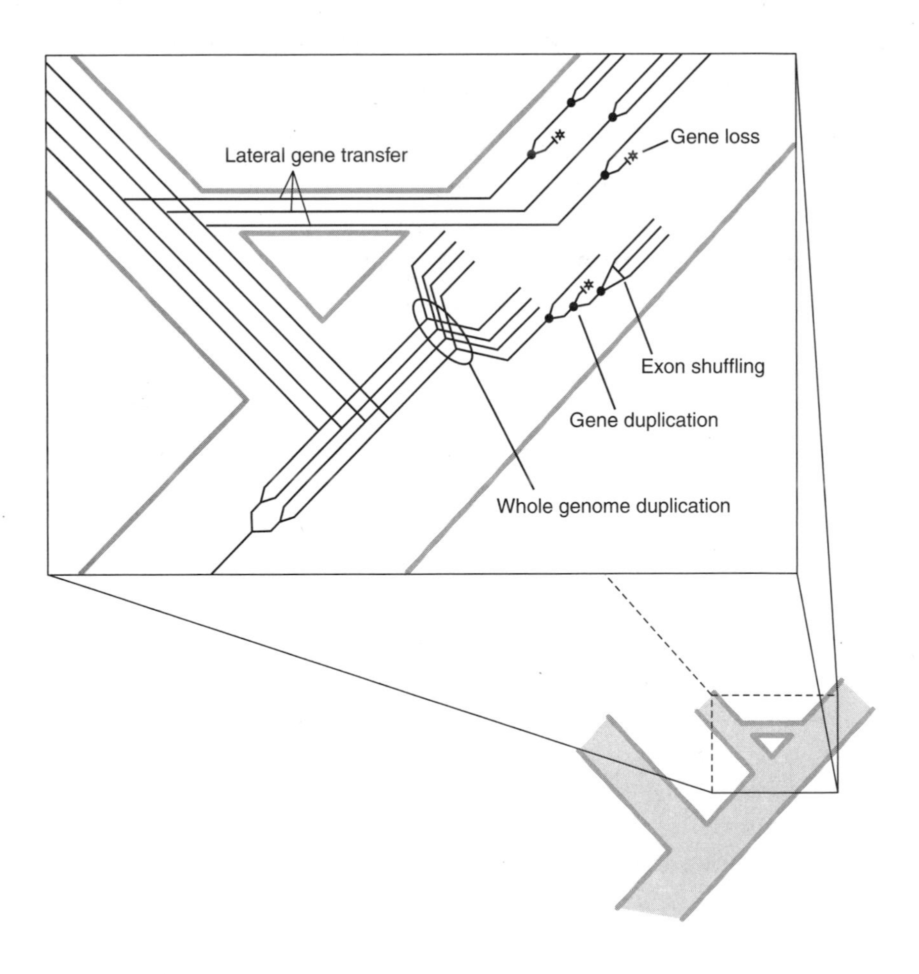

whole genomes is unrecoverable, owing to frequent change blurring the record of relatedness. What is recoverable provides the basis for a phylogenetic ordering of the diversity of genes. A first step entails using sequence alignment and phylogenetic analyses to identify the same genes in different species (orthologs) and to identify the phylogenetically related members of gene families (paralogs). These computationally intensive comparisons have begun. The clusters of orthologous groups (COGs) that have been identified are important for predicting the function of newly sequenced genes.[5] Using phylogenetic relationships of genes to predict gene functions, typically shared among COGs, is central in studies of the genetics of disease and genetic engineering in general by identifying appropriate orthologs for study and manipulation in model organisms in the lab.

Species are the primary units of evolutionary study for organismal lineages. Species denote the different kinds of organisms and are commonly viewed as reproductively isolated from each other. Reproductive isolation cannot be applied in considering asexual groups on which estimates of relative evolutionary distinctiveness are made instead. Approximately 1.5 million species have been identified and described by scientists to date. The number of undescribed species may be ten to a hundred times as large. Vertebrates are the most thoroughly surveyed group, followed by plants. Outside of these two groups our basic knowledge of the numbers of species, their geographic ranges and life history attributes, particular adaptations, and functional role in ecosystems is rather limited. Our knowledge of the different kinds and capabilities of viruses and bacteria is rudimentary, even though these groups are responsible for most infectious diseases of humans, livestock, and crop species. Similarly, accurate surveys of the planet's nematodes, mollusks, protozoans, algae, and fungi are lacking, despite the fact that such an inventory is a first step in resource management. We know a great deal more about the composition of the human genome than we do about the species composition of the Everglades or Manhattan, much less the earth.

A new biodiversity survey method is revealing a wealth of diverse microscopic life forms previously unimagined. "Culture-independent" polymerase chain reaction (CiPCR) obviates culturing of organisms in the lab. It works by extracting DNAs from soil or water samples, sequencing the DNA fragments, matching the overlapping ends of the fragments, and assembling the distinct linear DNA sequences for the various microscopic organisms swept up in the initial sample. Use of this approach has led to identification of many new lineages of eukaryotes so small that they are known as nanoeukaryotes—some being as small as bacteria. These are not young taxa, whose lineages are well known. Rather, some of them represent ancient groups and their inclusion in phylogenetic analyses is changing our understanding of the range of diversity and the branching pattern for the tree of life.[6] The CiPCR method has also more than doubled the number of named phyla for Bacteria and Archaea, even though most habitats remain unexplored with this method.

Individuals and species interact continually within communities and with the environment. These interactions are integral to the reproductive efforts of individuals and to the evolution of populations and species. Interacting communities together with their environments are known as ecosystems, and because these communities function collectively in cycling the raw materials for sustaining life, such as chemical elements and water, they constitute an important level of organization for the consideration of biological diversity. As life forms have evolved and their interactions have become more complex, ecosystems have diversified and increased in complexity.

Change within ecosystems, in turn, changes the selection pressures on individuals.

Ecosystems are generally delimited by the physical conditions of the environment, including water and soil characteristics, climate, altitude, and geographical locale. Location matters in description of ecosystems because local floras and faunas vary widely in composition as a result of the variable local history of species origins and dispersal. Ecosystems can range in scale from the global to the microscopic. Some of the primary types are: tundra, boreal forest, grassland chaparral, temperate forest, temperate rainforest, tropical forest, savannah, and desert. Within each of these there are a multitude of local ecosystems corresponding to local variation in precipitation, soil acidity and nutrient value, and elevation.

The Value of Biodiversity

Earth is rapidly losing biodiversity. Estimates of the numbers of species going extinct are in the range of 1,000 to 10,000 per year and rising.[7] There have been other major episodes of species extinction in the past, such as at the end of the Cambrian period 505 million years ago. In that extinction, climate and sea-level changes transformed habitats around the world. At the end of the Cretaceous period 65 million years ago, according to current evidence, one or more meteorites collided with the earth, sending huge clouds of debris into the biosphere, blocking sunlight, disrupting ecosystems, and cooling the climate. The current extinction episode, however, which equals or exceeds the others in its speed and taxonomic breadth, differs in its root cause. For the first time, large-scale extinction events are the result of the impact of one species—ours—as we multiply, disperse, and alter the earth's environments.

Should we be concerned about this loss? Yes, of course, and

few would seriously claim to be wholly unconcerned. The greater difficulty lies in motivating people and societies to act on those concerns and implement wise conservation policies. The difficulty arises, in part, because conservation of biodiversity requires some willingness to forgo short-term gains for longer-term benefits. Even though many of us profess concern for conservation of biodiversity, we are not often able to see its loss directly, and so we often fail to perceive it as an immediate or serious problem for our own lives. This perceived distance from any real consequences makes the issue an urgent one, and raising awareness of biodiversity's value ought to help. The reasons for valuing biological diversity are both economical and ethical.

Life's 3.8 billion-year history began and continues as a series of natural experiments. In every generation of every species, countless minor variations in genes, structure, function, and behavior are mixed and tested by natural selection for current suitability. The resulting varieties of form and function include adaptations at every level from genes, cells, and tissues to integrated whole organisms for the many activities of living. The list of activities, occurring at multiple levels, is pragmatic and familiar. It includes defense, attack, communication, surveillance, regulation, repair, reproduction, growth, and cooperation. Biodiversity presents a substantial repertoire of molecular and organismal abilities, many of potential use to people, that remain to be explored.

Many products derived from wild species are sold commercially. Prescription drugs, for example, represent a multi-billion dollar industry. They have become indispensable for many in relieving pain, bolstering immune systems, lowering high blood pressure, resolving infertility problems, and generally extending life spans. The active ingredients for over 75 percent of the 150 most commonly prescribed drugs in the United States originally came from organisms.[8]

Table 4.1 Drugs derived from wild plants

Plant	Distribution	Drug	Application
Willow	Worldwide	Aspirin	fever and pain
Cinchone	South America	Quinine	malaria
Rosy periwinkle	Madagascar	Vincristine/ Vinblastine	leukemia/Hodgkin's disease
Pacific yew	North America	Taxol	ovarian cancer
Opium poppy	Eurasia, Africa	Morphine	pain
Curare	South America	Tubocurarine	muscle relaxant
Snakeroot	India	Reserpine	hypertension
Foxglove	Eurasia, Africa	Digoxin	cardiac arrhythmia

Humans could not succeed in assembling such an array of useful compounds and medicines without access to the evolutionary fruits of biodiversity. The frequently used antibiotic tetracycline is isolated from a bacterium, and penicillin is derived from a fungus. Cyclosporine, another fungal-derived medication, has proven effective as an immune-system suppressant in successful heart and kidney transplants. Two drugs, vinblastine and vincristine, discovered in the rosy periwinkle *(Catharanthus roseus)* from Madagascar turn out to be effective in treating leukemia and Hodgkin's disease. Treatment with vincristine alone increases the survival rate for children with leukemia from 20 percent to 80 percent, though the cost of treatment makes it unavailable to many. Only a small proportion, less than 1 percent, of the named plants, fungi, and bacteria have been screened for useful pharmaceutical compounds. The wealth of biochemicals we employ and those we have yet to discover exist only because of the creative and refining actions of evolution via natural selection.

Bacteria adapted to living at extremely hot temperatures are known as "thermophiles." As noted above, some conduct routine enzymatic reactions, including replicating DNA, at boiling temperatures. Biologists have successfully learned to

Table 4.2 Features and applications of some Archaea and Bacteria living in extreme environments

Features	Applications
Thermophiles and Hyperthermophiles	
DNA polymerases	Copying DNA fragments
Lipases, pullulanases, proteases	Detergents
Amylases	Baking and brewing
Xylanases	Paper bleaching
Halophiles	
Bacteriorhodopsin	Optical switches and photocurrent generators
Lipids	Liposomes for drug delivery and cosmetics
g-Linoleic acid, b-carotene, cell extracts, e.g., *Spirulina* and *Dunaliella*	Health foods, dietary supplements
Psychrophiles	
Alkaline phosphatase	Molecular biology
Proteases, lipases, cellulases, amylases	Detergents
Polyunsaturated fatty acids	Food additives, dietary supplements
Ice nucleating proteins	Artificial snow, food industry, e.g., ice cream
Alkaliphiles and Acidophiles	
Proteases, cellulases, lipases, pullulanases	Detergents
Elastases, keritinases	Hair removal
Cyclodextrins	Mask odors, convert liquids to powders
Acidophiles	Paper making, waste treatment, solvents
Sulphur-oxidizing acidophiles	Recovery of metals and de-sulphurication of coal

Source: http://www.mediscover.net/Extremophiles.cfm

extract and use thermophilic polymerases in order to copy and sequence DNAs rapidly and at high temperatures in the lab. Much of the progress in molecular biology and its application to health and medicine has been accomplished with polymerases extracted from captive thermophilic bacteria. Billions of dollars have been made in sales of polymerases and sales of stock in the companies manufacturing them. There are other uses for bacteria adapted to life in extreme environments as well. For example, the protein bacteriorhodopsin, found in some halophiles (bacteria that can live in extremely salty environments), can convert light into energy and is used in making photocurrent generators. Sulphur-oxidizing enzymes found in some acidophiles is used to remove sulphur from coal, and enzymes from acidophilic bacteria are used as organic solvents.

Use of products from wild species is a longstanding tradition. Medieval scribes copied the Bible on goat or sheep skins using as ink the brown liquid ejected by cephalopods such as octopi at times of distress. For people living subsistence lifestyles, significant amounts of their food, medicines, wood fuel for cooking, heating, and housing materials are collected directly from organisms living in their natural habitats. For example, in Cameroon, wild-caught animals, including snails, caterpillars, wasp larvae, and other insects, constitute 70 percent or more of the animal protein that people in that country consume. Globally, more than 100 million tons of wild fish, crustaceans, and mollusks are taken annually for local consumption. About 80 percent of the world's humans rely on traditional medicines derived from plants and animals. In China and the Amazon basin, over 5,000 and 2,000 plant species, respectively, are known to be used for medicinal purposes.[9]

Wild species taken from natural environments and sold commercially represent another level of value in biodiversity. Primary natural products sold at market include construction timber, fuel wood, fish, shellfish, wild vegetables, wild fruits, wild game for meat and skins, feed for animals, wild beeswax, wild honey, fibers for clothing, and natural dyes. Another large category includes plant resins and gums used in varnishes, paper production, perfumes, medicines, embalming corpses, tanning hides, and thickening foods (including ice cream). Timber products from natural environments (as opposed to plantations) alone result in more than $120 billion in trade annually, though this is not sustainable.

Only a fraction of the wild species with potential for domestication have been developed. Local wild species or populations are often better suited to local conditions, having been selected for resistance to local pathogens and environmental conditions. Wild populations are also an important source for the genetic enhancement of existing domesticated populations of the same species. Crossing individuals from wild populations with domesticated forms may be useful in artificially selecting crop species for protein content, flavor, size, pest resistance, or extending environmental tolerances. It may be possible to transform annual crop species like corn into perennials by cross-breeding them with perennial relatives. Development of perennial corn crops would reduce production costs as well as the erosion that occurs while farmland is fallow. When invasive species become pests, wild species limiting their population sizes in their native ranges have been used successfully as natural biological control agents.[10]

The uses of biodiversity I have discussed thus far are based primarily on human consumption. Even more valuable is the role of biodiversity in maintaining functional ecosystems. Such ecosystem services include the protection of water and soil resources, the regulation of climate, and the cycling of nutrients and waste products. These complex operations require functional communities of species of organisms. Estimating the cost of these services is difficult and at some level arbitrary, much like estimating the costs of alternative economic policies. However, initial estimates suggest ecosystems services are worth trillions of dollars annually.[11]

Plant and algae species use sunlight to convert carbon dioxide into oxygen. In doing so, they create living tissues suffused with carbohydrates. These tissues provide the nutrient base sustaining virtually all animal products harvested by humans. Destruction of plant and algae communities, whether through logging, grazing, burning, or clearing for human development reduces the nutrient base supporting these animal communities. The vegetation of coastal estuaries is particularly important in supporting commercial fisheries and the noncommercial species on which they depend, and they are particularly endangered by human development.

Loss of plants and their communities also reduces uptake of carbon dioxide, whose excess leads to trapping of the sun's heat and global climate warming. Plants replenish the oxygen supply that most life forms require. Plant roots and the soil organisms they foster both help to aerate soils, increasing their ability to hold water. Deforestation, loss of vegetative cover, and loss of the associated communities of soil organisms has led to catastrophic flooding of low-lying agricultural areas in India, the Philippines, the midwestern United States, and elsewhere. Unabsorbed runoff carrying soil can kill freshwater and marine organisms. Heavy silt loads make river water undrinkable and its accumulation can reduce the navigability of waterways, destroy farmland, and prematurely fill reservoirs meant to produce electricity. However, events like these are usually seen by the public as local, sporadic tragedies without broader consequences.

Communities of bacteria and fungi are particularly important in successfully breaking down pollutants (such as pesticides and heavy metal pollutants) as well as in decomposing sewage and dead organisms. Their contribution is crucial to ecosystems; life without decomposition of the dead is unimaginable.

Biological diversity is also important in the economies of recreation and tourism, and some governments are beginning to understand the value of healthy ecosystems in that context. For example, a recent study by the Australian Productivity Commission reports that the Great Barrier Reef brings in over $4.2 billion U.S. dollars (USD) annually and employs about 48,000 people in the tourism sector. Combined with the value of recreational and commercial fishing on the reef, the value of biodiversity-related industries is approximately $5 billion USD per year in Australia.[12] In Botswana, international visitors spend over $315 million USD annually on wildlife tours, making this one of Botswana's most rapidly growing economic sectors. Ecotourism contributes more than three times as much to the gross domestic product (GDP) as agriculture.

The value of biodiversity and healthy environments for human well-being is widely known. Few would argue that ever-expanding human populations and declining numbers of wild species in smaller and more fragmented parcels of natural habitat are desirable. The understanding is also widespread, in both developed and developing regions of the world, that forests, fisheries, water, and other natural resources are finite and subject to human depletion. Understanding the value of natural environments and biodiversity, however, and acting to sustain them are two different things. Our consumption of natural resources is still accelerating. Most environmental degradation is incremental and invisible to the vast majority living in population centers. Further, the long-term value of natural resources rarely outweighs short-term economic gain, with its immediate payoff. Politicians often decide resource-use policies in favor of short-term benefits that can be delivered before the next election, and the public has little stomach for slowing resource consumption, especially when the long-term costs are hidden and deferred.

But deferring conservation of biodiversity and environments can have serious consequences. Jared Diamond chronicles the collapse of whole societies resulting, in part, from the failure to practice conservation of key natural resources, from Easter Islanders who vanquished the forests that provided their fishing canoes to Norse settlers of Greenland who depleted the sparse soils by overgrazing.[13]

Why is it so difficult to conserve shared resources for the

long-term public good? There are many reasons, including the trade-offs between self-interests and group-interests and the more immediate benefits of short-term gains noted above. Another reason often cited is the lack of a strong conservation ethic. Western religious traditions have had a significant historical role in the development of ethics, identifying socially unacceptable behaviors, and those traditions have not stressed conservation and stewardship of the earth's natural resources. Although some religious teachings are consistent, in hindsight, with a concern for the environment, an ethic of commitment to care for the environment has been weak at best. The attitudes of dominion over nature and entitlement to its resources to satisfy human needs have been much more prominent in religious cultures. Arguably, the ethics stemming from religious traditions reflect an earlier time when resource limitation was little known.

Does an evolutionary view shed any light on the ethics of conservation? From an evolutionary point of view, ethics may be seen as a form of social contract with benefits, in terms of survival and reproduction, for both individuals and the social groups, especially kin groups, to which they belong. If the social group includes lots of kin, ethical behavior among them, promoting their shared gene pool, may be favored over time. As I discuss in Chapter 5, the capacity for ethical social behavior is one of the human traits that underlie the development of societies and religions. A variety of religious traditions have had success in promoting ethical behavior, regardless of the relatedness of individuals, by linking ethical behavior to personal fulfillment. Those who practice ethical behaviors benefiting the larger group may find salvation, greater self-esteem, and social acceptance.

Might existing religions develop a stronger conservation ethic in the near future, one that is widely embraced? This seems unlikely, given current attitudes about the environment.

Loyal Rue has outlined a possible future scenario in which severe environmental degradation does lead to large-scale collapse of human societies. He supposes that the survivors of the collapse will have a much deeper understanding of our need to maintain biologically diverse and healthy environments. Over time, the survivors develop religious cultures which make the same links between self-esteem and ethical behavior that traditional religions have made, but they do so with little or no distinction between God and nature. In his words, "God is naturalized and Nature is divinized."[14] This is, of course, extreme speculation regarding cultural evolution. But he does call attention to the real potential for societal collapse and a current trend within some religious cultures to develop ethics more focused on the natural than on the supernatural.

In line with a focus on naturalism, there is a case to be made that the earth's diverse species are inherently worthy of conservation on ethical grounds. All life forms share a common evolutionary history. This shared ancestry defines life as a single family. It is a highly extended and diverse family, but a family nonetheless, and this genealogical connection supports an ethical concern for biodiversity conservation. But an ethical concern for other species will be perceived differently, if at all, by different individuals, and it is likely to be felt more strongly for species more closely related to humans or species perceived as charismatic. Perhaps a more compelling argument focuses on conservation of biodiversity resources for future generations of humans as an ethical motivation. Certainly, long-term conservation has the potential to enhance human reproduction and survival. But, historically, this has led more often to fighting over control of resources than to cooperation in a shared ethic of conservation. None of these justifications for conservation will last long where they conflict with perceived human needs.

The pragmatic, existing approach to conservation involves

legislation for sustainable use and maintenance of shared resources. This requires long-term thinking, and some sacrifice of short-term economic gain. Long-term planning and sacrifice for generations in the distant future may seem antithetical to our biological imperative to reproduce and provide for our own offspring. But the lesson we learn from assessing the value of biodiversity and healthy environments is that their conservation yields benefits for both the short and long term. We have also seen that the distance between short- and long-term time frames can collapse overnight. Using our capacity for learning, planning, and ethical behavior, developed over evolutionary time, is a key component in solving our conservation problems.

Applications of Phylogenetics

Describing and Organizing Diversity

A primary task in any attempt to understand and manage resources is taking inventory. If we do not know what species exist and where they are found, we cannot begin to understand their unique attributes, their roles in communities, or their value for humans. The basic task of inventorying the earth's species or kinds requires exploration of all regions for the kinds of organisms present, finding representative individuals of those different kinds and if possible collecting them or sampling genetic materials, describing their characteristic features, and comparing them in phylogenetic analyses with those of other organisms.

The conventions of species recognition vary among groups of organisms. For example, among sexually reproducing vertebrates, reproductive isolation from other groups is the defining property. But (as previously noted) this particular definition cannot be applied to asexually reproducing organisms, such as bacteria, where species are generally defined as organisms sharing unique phenotypic, DNA sequence, and ecological characteristics. Even for sexually reproducing organisms, knowledge is often lacking about whether particular individuals represent reproductively isolated species or not. In these cases, morphological, molecular sequence, and ecological differences, often apparent as shared derived traits for groups of individuals in phylogenetic analyses, can serve as a guide. Thus, the phylogenetic analyses mentioned above, as a basic part of inventorying, can provide a measure of distinctiveness among groups of organisms and the means for quantifying the process of recognizing and naming species. Phylogenies also identify species' closest relatives, and can be used to estimate when different groups last shared a common ancestor.

The formal task of organizing all species into a hierarchical system of ranked categories is known as classification or taxonomy. Taxonomy dates back at least to Linnaeus' implementation of a binomial system in the 1750s, and the naming of organisms and ordering them into groups is as old as human language. Today, taxonomists are evolutionary biologists with comprehensive knowledge about diverse taxa. Modern classifications of organisms seek to place the diversity of known forms into a hierarchy of species and higher-level taxa reflecting their phylogenetic relationships.

Species that are at risk of extinction are units of immediate conservation concern. Thus, determining species status and resolving taxonomic uncertainties by means of phylogenetics is a primary application of evolutionary theory in conservation efforts. In some instances, widespread taxa that are uniform in outward appearance may harbor distinct species, identifiable based on molecular sequence differences. This was found to be the case among tuataras, a group of New Zealand reptiles. Tuataras are the only surviving lineage of a formerly species-rich group known as Rhyncocephalians. One of their unusual

Figure 4.2 Three cases of applying phylogeny to help resolve taxonomic uncertainties, showing (left to right) phylogenetic trees, the conventional taxonomy for the individuals analyzed, and the phylogenetically supported taxonomy for those same individuals. (a) Phylogeny for Australasian teals (waterfowl) based on mitochondrial DNA supporting taxonomic recognition of three species of "New Zealand" teals instead of the single species traditionally recognized *(Anas chlorotis)*. (b) Phylogeny for elephants based on nuclear and mitochondrial DNA supporting recognition of two species in Africa instead of one. (c) Phylogeny for kites (raptors) in the genus *Milvus* based on mitochondrial DNA supporting dissolution of the Cape Verde kite as a distinctive species and recognition of a new species that had been considered a group within *M. migrans*.

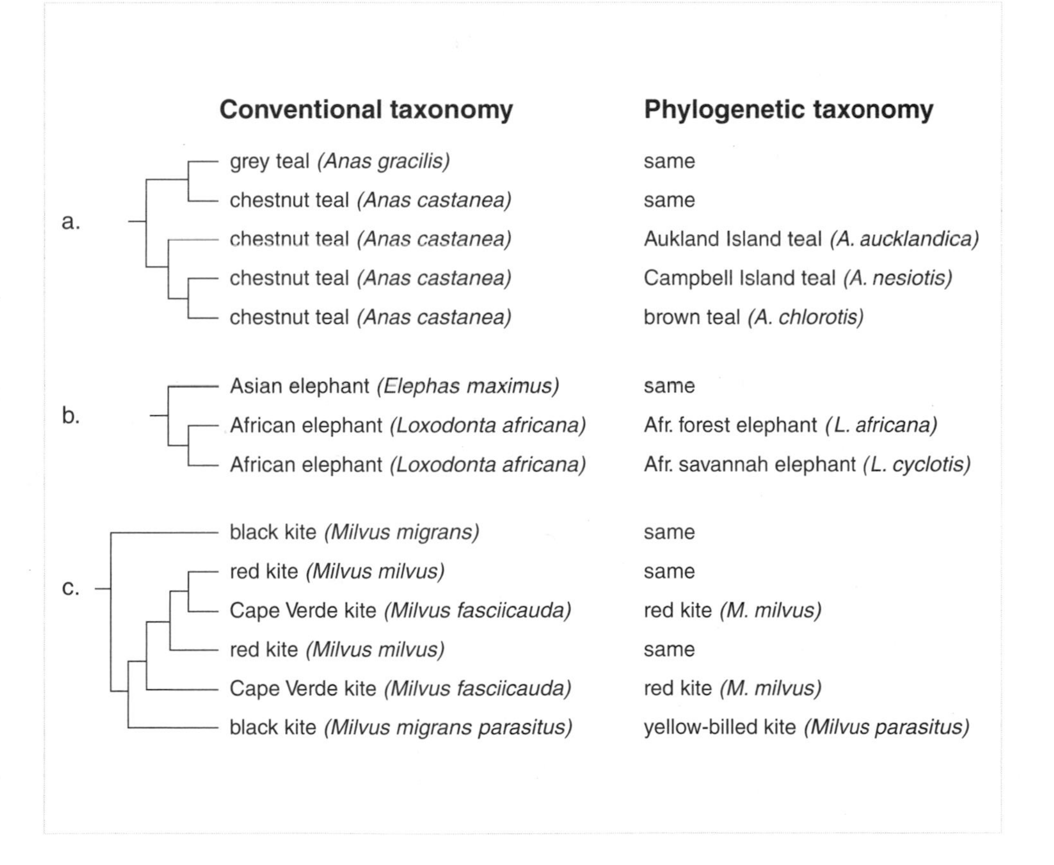

traits is a light-sensitive pineal eye on the top of the head, covered by scales in adults but exposed in newborns. Molecular analyses of this relict group revealed two distinct species, one of which *(Sphenodon guntheri)* was at risk of extinction. This species could be found only on a four-hectare island in Cook Strait. Its population was small, and threatened by predatory feral cats and rats.[15] Divergent species, previously unrecognized, have also been implicated in phylogenetic analyses of DNA in Asian soft-shelled turtles, right whales, and Australasian teals.[16] DNA sequence analyses of free-ranging African elephants have revealed significant differentiation between populations inhabiting forest versus savannah habitats. The genetic difference between populations within what had been considered a single species, the African elephant *(Loxodonta africana),* is about 58 percent as great as the difference between the two different elephant genera on different continents, represented by *Loxodonta africana* and *Elephas maximus,* the Indian elephant. Phylogenetic analyses of both nuclear and mitochondrial genes agree, and we now recognize two elephant species in Africa, *L. africana* (found primarily in forest habitats) and *L. cyclotis* (found in savannah habitats).[17]

The phylogenetic approach can also result in reducing species or subspecies numbers where differentiation is lacking. Fewer species or subspecies means less competition for scarce conservation resources. For example, molecular sequence analyses showed that the colonial pocket gopher from Georgia was indistinguishable from the common pocket gopher, and its former status as endangered has been revised. Though molecular phylogenetic studies of birds are revealing some distinctive taxa that have not yet been named and recognized, other studies are showing some named avian subspecies to lack significant genetic differentiation for mitochondrial markers and that some conventionally recognized species may not warrant species status.[18] Phylogenetic analyses have raised the possibility that endangered red wolves from eastern North America are hybrids between grey wolves and coyotes, though the issue is not yet resolved.[19] These and other examples reveal the need for more systematic analyses in identifying taxa and setting conservation priorities.

A growing number of wild species of animals and plants are being lost through hybridization with introduced or invasive species.[20] It will be increasingly important to identify taxa of hybrid origin, to know whether they arise from natural or human-induced hybrid origins, and to determine whether any purebred populations remain. Conservation policy for hybrids will vary case by case, though the first step is having good information about its occurrence and history. Hybridization history is best examined with genetic haplotype data (unique patterns of alleles or DNA sequence variation) characterizing both individuals and populations, with hybrids showing various degrees of admixture of the traits of purebred populations.[21] The methods are based on heritability of genetic traits and are reasonably seen as an application of evolutionary theory to conservation.

Biodiversity discovery and classification are basic scientific tasks in determining informed conservation practices. Just as genomic-based studies of human health will depend on the catalogue of mapped and annotated genes from entire genomes, studies of biodiversity health rest on comprehensive efforts to discover, classify, and annotate the natural histories for species of all kinds. Collections of specimens in research museums, herbaria, and microbiology labs around the world are crucial in this discovery and classification of species. These collections provide the reference standards for identification. They encompass the variation, both within and among species, that is the foundation of phylogenetic analyses. Each specimen and its label provide information on geographic and ecological distributions that can be compared over time. Distribution changes can be dramatic, especially in light of climatic changes such as that which we have been experiencing in recent decades. For many groups of organisms, archived specimens can also provide DNA, and thus constitute irreplaceable resources for scientists unable to travel to the organism's native habitat for financial or political reasons. In some cases, the DNA in a specimen can never be replaced because the species is now extinct. For these reasons, the expertise of taxonomists and training programs for students interested in biodiversity and taxonomy are crucial if we are to progress with our inventory of the earth's species and understanding of biodiversity.

Identifying Conservation Priorities

The goal of conservation biology is to sustain biodiversity. This can be accomplished through the protection of species and habitats. However, prioritizing among all the species and habitats in targeting conservation efforts is not easy. Rather than focusing solely on particular species or habitats, there is growing consensus among biologists that priority should be given to preserve the processes that support the origins and maintenance of biodiversity.[22] Focusing on the evolutionary and ecological processes sustaining life puts the focus where it should be, on the long-term persistence of biological diversity. Three steps can be outlined in identifying conservation priorities, and these correspond roughly to consideration of the past, present, and future of evolutionary history.

A first step is phylogenetic analysis, as noted in the previous section. Phylogeny illuminates past evolutionary history, identifying genealogy of taxa as well as the extent of differentiation among them. This helps in the recognition of species as characterized by distinctiveness and, if relevant, reproductive isolation. Phylogenetic tree branch lengths denote approximate amounts of diversity in traits among taxa, and this provides guidance in conserving as much of the diversity in traits inherited from the past as possible. For example, the two tuatara species discussed, when placed in a phylogenetic analysis with all extant Reptilia, are shown to represent the only survivors of a distinctive lineage that diverged from the ancestors of extant lizards and snakes about 230 million years ago.[23] Knowing their early phylogenetic divergence among reptiles and their lack of close relatives indicates their value in attempts to conserve diversity in both genetic, morphological, and behavioral traits. The coelacanth *(Latimeria chalumnae),* a primitive bony fish known from a few locations in the Indian Ocean, is another example of an uncommon, phylogenetically distinctive species whose conservation promotes maintenance of extensive genetic and morphological diversity not found elsewhere.

Phylogenetic analysis for individuals from different geographic regions within a species range is known as phylogeography. Combined with a variety of measures of genetic similarity, phylogeography can provide information on the de-

Figure 4.3 A temporal framework for setting conservation priorities.

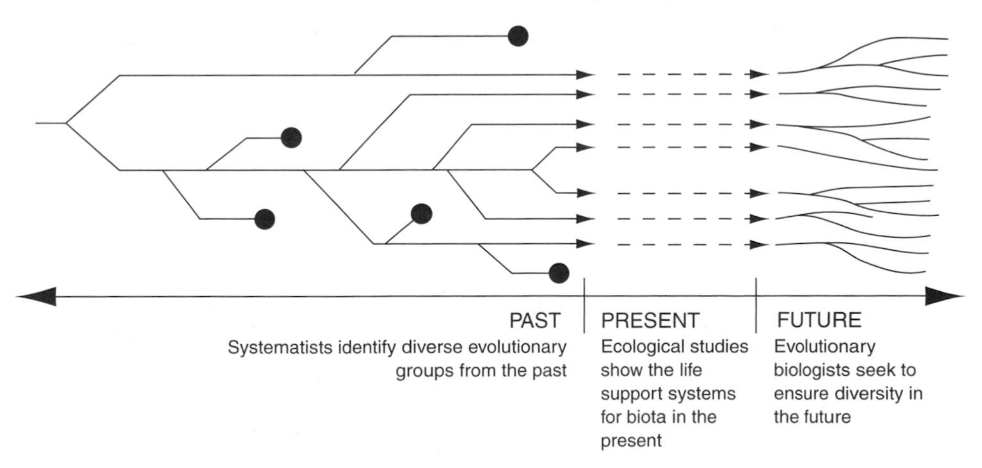

gree of genetic diversity within and between populations needed by conservation biologists to estimate relative population health. Low levels of genetic diversity can decrease the ability of species to respond to changing environments, and outbreeding efforts between conspecific populations or recently diverged species can add needed genetic diversity. Phylogeography can be useful in identifying candidate breeders for outbreeding efforts aimed at enhancing genetic diversity where needed. Augmenting genetic diversity in wild populations by introducing individuals from captive breeding projects must be done with caution. Captive populations can evolve adaptations to their environments just as wild populations develop adaptations to theirs, and the adaptations of captive breeding populations are unlikely to be advantageous in the wild.[24] Phylogeography and similarity measures also reveal the relative amounts of gene flow among geographic populations. This can inform efforts to conserve genetic diversity within species and to identify and protect important geographic regions and habitat corridors where gene flow occurs as a result of dispersal of individuals.

A second step in identifying conservation units and priorities includes ecological studies of the interactions among individuals within species communities and their interactions with their environments. Ecological studies illuminate current evolutionary processes for organisms and show which resources and species interactions are crucial for maintenance of healthy populations and ecosystems. But it is not always easy to delineate the components or determine the area and population sizes of an ecosystem. Our greatest losses in biodiversity are a result of loss of habitat, and conservation of inclusive ecosystems with viable habitats is the key to sustaining taxa over the long term. If we work to conserve ecosystems, we conserve ecological processes as a result, and in the final outcome, taxa are conserved as well. For example, predation is an important ecosystems process, and efforts to conserve ecosystems encompass conservation of predators, such as panthers and wolves, along with populations of their prey species and the habitats sustaining them.

A third step in identification of conservation units and priorities must be the consideration of evolutionary potential. The evolution of novel, future biodiversity is needed to keep pace with continual environmental change and the novel selection pressures that will be imposed on taxa and ecosystems. Thus, successful conservation will seek to protect existing biodiversity as well as the resources and processes important to the origin of future biodiversity.

One way to enhance future biodiversity is to identify taxa that are members of species-rich groups with a demonstrated capability for rapid evolution, and to identify those ecosystems and geographic regions spawning them. In contrast, ancient lineages with few close relatives (living fossils) may not be strong candidates for generating future biodiversity.[25] Evolutionarily dynamic lineages and the geographic regions most conducive to them can be identified phylogenetically. Species radiations, seen on phylogenetic trees as clusters of branching events in relatively short periods of time, are indicative of high levels of evolutionary potential. Identification of geographic regions, or ecosystems, where species radiations have occurred in multiple groups of taxa indicate the fertility of these regions or ecosystems as growth chambers for biodiversity. A caveat is that evolutionary potential for lineages is dependent on opportunities available for differentiation. Such opportunities are difficult to predict. Features to conserve that are often associated with rapid diversification among taxa and local species-richness are high levels of genetic diversity within and among populations and heterogeneity and connectivity of their habitats. On a global scale, coral reefs and tropical rainforests are two such important regions and ecosystems for generating biodiversity. However, conservation efforts applied locally within regions and habitats of all types can benefit from this approach. For example, cichlid fishes in eastern Africa and honeycreepers in Hawaii are examples of species-rich lineages with capacity for rapid change, and conservation efforts to maintain them may preserve sources for future evolution.

Bioprospecting

Bioprospecting involves the discovery and development of natural resources, particularly biologically active compounds for medicinal uses and genes encoding functions potentially useful in agriculture or animal breeding.

Because these resources have evolved over long periods of time, often as adaptations, and because traits tend to be shared among relatives, phylogenetics is reasonably applied in bioprospecting. Once identified in one species, potentially useful compounds and the genes encoding them are more likely to be found by examining close relatives than by randomly searching among organisms or mixing chemicals off the shelf. Using phylogenies to guide efforts to locate useful genetic and biochemical resources is only the first step in bringing such products to market. Isolating, testing, and augmenting the natural products are the more labor intensive and difficult tasks. Comparative analyses can also reveal the diagnostic traits of species known to be useful, but that are easily confused with similar species.

For example, a phylogenetic approach has been useful in the (ongoing) development of the anti-HIV medication calanolide A, isolated from a rare tropical tree, *Calophyllum lanigerum* variety *austrocoriaceum.* Calanolide A prevents HIV from entering the nuclei of healthy human white blood cells.[26] This was first discovered based on work with extracts from leaves of a specimen collected in Borneo. Comparative phylogenetic analysis using herbaria specimens revealed the unique traits for this tree, enabling researchers to find additional specimens and tell them apart from closely related species lacking the compound.

Conservation Forensics

If we are to succeed in slowing biodiversity loss and conserving a significant fraction of the current survivors, we will need greater resolve and better legal protection for habitats and taxa. Even though most species are not the intended quarry of human hunters, an ecosystems view tells us that protection of fish, game, and pet-trade species is important in healthy functional environments and that protection of their populations will benefit those of other species as well.

Commercial fishing, whether for meat, caviar, or fins for soup is not sustainable at current levels. Items such as rhinoceros horns, elephant tusks, marine turtle shells, corals, and bear gallbladders, which are used variously as aphrodisiacs, talismans, art, or medicines, increase in black-market value as they become more rare. The same is true in the pet trade: wild populations of tropical fish, amphibians, reptiles, and birds are increasingly tapped and at risk. There are some regulations on the numbers and species of wildlife that can be harvested; however, unless these regulations can be enforced, they have no effect. Fish and game products are often processed beyond recognition before reaching open markets. Geographic origins are frequently not known, and in some cases protected species are difficult to distinguish from legally harvested ones.

Application of genetic markers in evolutionary analyses will be increasingly important in enforcement of conservation regulations. These applications fit into four categories: phylogenetic analysis of DNA to identify species, phylogeographical analyses of DNA to identify geographic origins, similarity matching of hypervariable DNA "fingerprints," and sex determination. The first three of these applications are based on the core principle of descent with modification and use comparative methods to infer the history of relatedness. If the species and origin of a specimen can be determined definitively, prosecution of those who violate conservation laws is simplified. Sex determination is useful in testing compliance with gender-specific harvesting of game species. Application of these methods is only beginning, though the basic concepts have been demonstrated and the methods and databases are being further developed for broader applications.

For example, phylogenetic analyses of mitochondrial DNAs extracted from meat purchased at Japanese and Korean whale meat markets focused attention on the lack of compli-

Figure 4.4 Phylogeny based on mitochondrial DNA for select whales and dolphins and for meat samples of unknown identity purchased in commercial markets ("? sample"). The purpose of the phylogenetic analysis was to see if protected whale and dolphin species were being sold illegally.

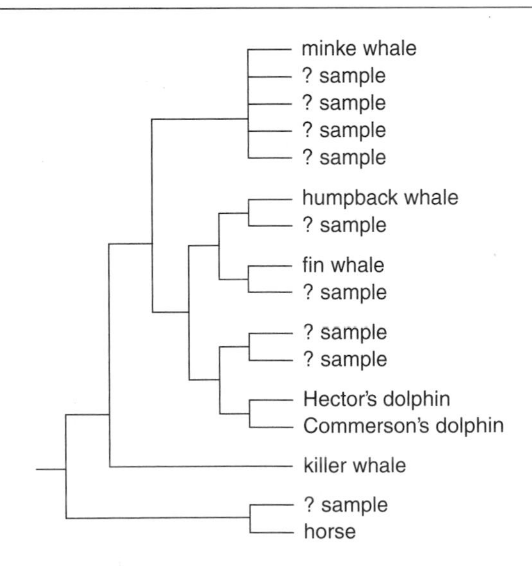

ance with whaling regulations by identifying conclusively the sources of intentionally mislabeled meat. Whale meat being sold originated not just from minke whales, which could be legally hunted, but from humpback, fin, blue, and sei whales as well, all of which are protected.[27] Some of the meat turned out to be from dolphin, horses, and sheep too. The same approach has shown the presence of tiger products in Asian medicines and identified caviar taken illegally from protected sturgeon species.

Phylogeographic analyses have been used to show that a group of illegally captured chimpanzees originated from Uganda, based on existing sequence data sets for chimpanzee populations. Subsequently, efforts were made to increase surveillance, and the confiscated chimps were returned to their native locale.[28]

In another application, a suspected deer poacher in Florida claimed that fresh blood on his clothing was from a young cow he had butchered; however, analyses of the blood showed it came from a deer killed out of season, and further that it came, also illegally, from a female. Many challenges exist in implementing these approaches. These include developing population genetic databases for identifying geographic origins for specimens, securing funds for the necessary sampling and lab work involved in enforcement, making sure the analyses are sound and well documented for use in court. Despite these obstacles, evolutionary forensic methods are certain to help in enforcing conservation regulations.

Applications of Natural History Knowledge

A primary concern of conservation geneticists is measuring and managing the genetic diversity of species and populations at risk. Low levels of genetic diversity limit the ability of populations to evolve and respond to environmental change. This can be perilous, as natural populations face continual change in the distributions and armaments of pathogens, parasites, competitors, and predators, as well as human-induced change in climate, pollutants, and pesticides. Without genetic variation in populations of individuals, natural selection has little raw material on which to act. Low genetic variation is also correlated with small populations and inbreeding. Inbreeding often leads to reduced productivity and defects including abnormal morphology and poor-quality sperm or eggs. Crop plants have often been highly inbred, and as a result, they are more susceptible to attack from pathogens. For example, the fungal attack on potato crops in Ireland during the 1840s and the bacterial infection of Florida citrus trees with canker during the 1980s are both linked to low levels of genetic diversity. Inbreeding and genetic diversity loss have been implicated in population declines or extinctions for wild taxa including heath hens, prairie chickens, Florida panthers, and bighorn sheep, among others.

How is our understanding of evolutionary processes used to conserve biological diversity? One way is to enhance the evolutionary processes that give rise to genetic variation within populations, and this can be done by using knowledge of species' natural histories. The evolution and maintenance of genetic diversity depends on such factors as the size of populations, the frequency of interbreeding among populations, the age of individuals at first breeding, the rate of reproduction, and local differences in environments and adaptations to those differences, if any. If populations are small and relatively isolated from others (having reduced gene flow), transplanting individuals with variant genes between populations may boost genetic diversity. Creation of habitat corridors connecting currently separated populations may accomplish the same thing. These practices are useful in cases where populations have been isolated as a result of human activities. Management to increase population size can increase population genetic diversity as well. If some populations have unique capabilities or adaptations relative to others, efforts may be made to conserve that variation through selective mating programs. Captive breeding efforts may benefit from broad-scale genotyping of individuals and intentionally interbreeding divergent genotypes. Captive breeding populations should not be so small that they suffer from inbreeding and reduced genetic variation themselves. Occasional introduction of wild individuals into captive breeding stock promotes outbreeding and may reduce the real problem of adaptation to captivity. Some species have a slow reproduction rate or breed at an advanced age, such as California condors, which reproduce at age seven or so. Those taxa should be ranked as higher priority for captive breeding and release efforts, because recovery of populations in the wild will take longer.

Evolutionary studies of populations of fruit flies, yeast, mustard plants, or other model organisms in the lab have been used to assess many aspects of genetic diversity. These provide essential background knowledge for conservation biologists, even though the details of life history differ among taxa. For example, model organism studies have been useful in learning about the rates of genetic diversity loss and recovery in populations of varying size, productivity differences among individuals differing in heterozygosity (variation among the two copies for each gene), the effects of skewed sex ratios and alternative mating systems, and the effects of novel pathogens in populations of variable size and heterozygosity, to name just a few. A great deal of population genetics theory, a key component of evolutionary biology, is central to informing the theory and practice of conservation biology.[29]

Current biodiversity losses are driven by social and political forces, and conservation biologists' priorities generally take a back seat. There have been, however, some successes. Considering one of them provides an example of how an understanding of a species ecology and evolutionary history is relevant to conservation efforts.

Peregrine falcon pairs have been raising their young on the cliffs of Lundy Island, off the coast of southwest England, at least since 1243. On Lundy Island and elsewhere, successive generations of peregrines are well known for returning to the same nesting sites year after year, and when they began disappearing from these traditional breeding sites in the 1940s and 1950s in Europe and North America, naturalists became concerned. This was the beginning of one of the first and relatively few successful conservation efforts. Painstaking research over the course of ten years revealed that reduced productivity and population declines were the result of thin-shelled eggs breaking during incubation. The thin shells were caused by the pesticide DDT and its derivative DDE, which disrupted calcium metabolism. Peregrines eat birds that feed on insects and plant seeds, and the insects and seeds carry pesticides initially sprayed on crops by humans to control pests. The preferred prey of peregrine falcons are often long-distance migrants, including songbirds, shorebirds, and waterfowl exposed to pesticides on their wintering grounds. Because peregrine falcons are predators at the top of the food chain, DDT concentrations were higher in them than in their prey. Bald eagles, osprey, and brown pelicans also accumulated pesticides from their prey of choice, fish.

In response to the population declines and understanding of the cause, captive breeding and release programs were initiated, population monitoring studies were intensified, and DDT use was banned in many areas in the 1970s. Though not without controversy or setbacks, some populations of peregrines have been restored, and new breeding sites have been adopted, including some office buildings towering above urban canyons, home to swallows, swifts, and pigeons.

At each step along the way, biologists' knowledge of the life-history traits of peregrines—their abilities and behaviors—has been vital in devising appropriate plans. Further, this requisite understanding of their biology, from physiology to feeding habits and migratory behavior, is most comprehensive when placed in the context of their evolutionary history and the habits and constraints that are the result of thousands and millions of years of common descent. Captive breeding efforts only succeed if the individuals released are competent in the wild and able to survive on their own. During the first few weeks of life, hatchlings of some bird species "imprint" on whomever is feeding them (usually their parents) and learn to identify with them and trust them. Thus, biologists know to avoid having captive-reared peregrines imprint on humans. Wild birds, particularly migrants, fare better if they avoid humans. In particular, they need to identify with other peregrines in order to mate successfully. This behavior was first documented by Konrad Lorenz, working with graylag geese in the 1930s. Realizing that this evolved and heritable imprinting behavior exists, biologists fed captive-reared peregrine falcon hatchlings with hand puppets painted to look like adult peregrines and kept themselves hidden. Similarly, biologists sought to replicate the conditions under which peregrines had evolved in considering their diets, release sites, and in getting females to lay a second clutch of eggs following "loss" of the first set, taken for captive rearing. Effective planning by conservation biologists requires an understanding of species life histories as evolutionary constraints on species habits.

We have considered applications of knowledge regarding species' life history and evolution for conservation. Now let's consider how our understanding of the evolution and function of ecosystems is used in conservation of biodiversity. Ecosystems include the interactions of organisms among themselves and with the environment. Understanding the evolutionary history of particular ecosystems entails understanding the life histories of species and their evolutionary effects on each other. Species are said to coevolve if they exert enough selective force to provoke change in each other. Timing of reproduction for many predators is often locally tuned to slightly precede that of their favored prey. Plants and pollinators often exert selective pressure on each other for matching structures, search images, and timing of availability. The variable bill shapes and sizes in hummingbirds are generally well matched with the shapes and sizes of the flowers from which they extract nectar and for which they provide pollen dispersal.[30]

The coevolved, integrated nature of species life histories and ecological roles supports the premise that conservation of the processes of ecosystems matters. This is most effectively accomplished by the conservation of large areas, which encompass the full range of local habitat types that will benefit the full range of species in that area. Conservation efforts in crisis mode, aimed at single, well-known species, are often necessary, but ultimately a broader approach impacting more species is needed. Toward this end, habitat fragmentation should be minimized to reduce barriers to movement as individuals seek to reproduce and as species interact with each other in the cycling of energy and nutrients. Biologists will use their background knowledge to rank areas for priority in conservation based on the distinctiveness and endangerment of particular species and their roles in ecosystems function.

Knowledge of coevolution and ecosystems function also allows biologists to identify and manage certain keystone species and resources that are of greater importance to the survival of more individuals and species within an ecosystem than are others. Conservation of keystone species and resources is justified as a high priority because of their relatively high impact on biodiversity and ecosystem functioning. The Samoan flying fox *(Pteropus samoensis),* a fruit-eating bat with a wingspan of 2.5 feet, provides an example of a keystone species. It is the primary pollinator and seed disperser for numerous endemic plants in the Samoan and Fijian archipelagos of the South Pacific, including some economically important and ecologically dominant tree species.[31] Loss of this key pollinator could have dramatic negative consequences on forest regeneration and healthy outbreeding among plant populations. The sea otter *(Enhydra lutris)* is another example of a keystone species. Observed regional extinctions of sea otters led to population increase in sea urchins, their favored prey, and this led to overgrazing and loss of kelp and algae forests, which led to declines in fish populations dependent on the kelp and algae.[32] As a related example, loss of large predators and diminished biodiversity have contributed to the spread of Lyme disease in eastern North America by facilitating population growth and range expansion for the white-footed mouse *(Peromyscus leucopus).* This small mammal is a primary host of the bacterial pathogen *(Borrelia burgdorferi)* picked up by ixodid ticks having their annual blood meal. These ticks are then able to transmit the pathogen to humans.[33] Greater species diversity and population sizes of hosts that are less susceptible to infection with the bacterial pathogen would favor reduced Lyme disease in humans.

Most biological communities experience gradual, local changes in species composition and physical characteristics following disturbance. Knowledge of ecological succession among species within healthy ecosystems can be used to help restore or maintain the diversity of species over time, if needed. Ecological succession is the turnover among species within communities that results from progressive modification of the physical environment. Overgrazed grasslands and overcut forests are likely to have lost their rare late-successional species, and areas may be managed to allow further succession among species. Alternatively, habitats may require periodic burning or other disturbance to restart the successional process.

Biologists have long noted that species-rich ecosystems tend to be more stable and productive over time than are species-poor ecosystems. Direct testing of the extent to which biodiversity serves as insurance against ecosystem change is still needed;[34] however, when environments change and taxa go extinct, there are more candidate taxa to carry on particular roles in species-rich ecosystems compared to species-poor ecosystems. If one species of pollinator, predator, prey, or shelter provider goes extinct, there are more likely to be others to fill the role in species-rich ecosystems. Further, diversity tends to beget more diversity. More plant species with variable timing in reproductive readiness, flower shape, and scent create more opportunities for divergent insect or avian pollinators (or vice versa). More soil invertebrate taxa, differing in life histories, provide food in more habitats and seasons for more specialist predatory insects, sustaining more rodent insectivores, which support more predator species, and so on. In this broad view, biodiversity makes ecosystems more stable, especially in times of adversity. This will be a lesson to heed as human populations continue to expand.

5

EVOLUTIONARY METAPHOR IN HUMAN CULTURE

In the beginning, God said, "Let there be light." An optimist might respond, "Well, nothing else existed yet, but at least you could see it better." Genesis is the first book in the compendium of books written between 1000 BCE and 350 BCE known as Hebrew scriptures or the Old Testament of the Bible. It turns out that light is often used as a metaphor for knowledge or wisdom in scriptures, and when light is read as a metaphor in this passage, it makes more sense to our contemporary minds and even gains some poetic appeal. We cannot be certain, but it seems reasonable to assume that even 3,000 years ago readers or listeners would not have missed the difficulty of a strictly literal view in which light appears before light sources appear.

Metaphor is a useful device, in science as elsewhere, for communicating ideas or events. When the metaphor is mistaken for direct description, however, our understanding is diminished. If I say, metaphorically, that blood draws sharks like moths to a flame, it would be incorrect to assume a common mechanism of attraction. One involves attraction to odor, the other attraction to light. Similarly, saying that an eight-year-old boy is "growing like a weed" should not be taken to mean that he requires dirt and carbon dioxide or that he will produce oxygen.

A similar distinction should be made between the evolution of biological entities and the metaphorical evolution of differences among human cultures, such as differences among languages and religions. Biological entities evolve because of the differential reproductive success of individuals varying in their heritable traits (natural selection) and the influence of chance events. Biological evolution requires heritable variation in traits, and DNA is the only material directly inherited across generations. One can test the extent to which particular traits are heritable or not by observing the fate of parental traits in their progeny.

The *capacity* for human cultural development, including verbal communication and development of ethics, does evolve biologically, along with brain morphology, physiology, mental acuity, and aspects of social behavior. Many basic features of language construction, including features of syntax—the ordering and relationships among nouns, verbs, and adjectives—are shared by nearly all languages, and these have evolved in the biological sense to the extent that they are constrained by brain function. Language as a cognitive structure has been shaped by biological evolutionary processes over time, and can be fruitfully studied as part of the evolution and function of a complex body organ. Similarly, certain basic features of social behavior, including the capacity to develop and embrace a system of ethics, have been molded by biological evolution. Ethics are reasonably considered as a form of social contract with benefits for both individuals and the social groups to which they belong. The capacity for ethical social behavior is at least one of the human traits that underlie the development of religion. The cultural innovations resulting from this enhanced biological capacity for culture have had great impact on human evolution and to some extent on the differential success of human populations.

Thus, there is a strong connection between organismal evolution and cultural development; however, evolution for particular cultural features remains a metaphor. Particular changes in the small-scale details of cultural practices stem from differences in teaching, learning, imitation, and innovation as well as chance events. There is no gene constraining anyone to drink sake instead of wine, speak German instead of Yupik, or to believe in one, three, or no gods at all. The significance of this is that the details of cultures do not change over time as a result of the same mechanisms as genes and genetically encoded organismal traits, and it would be misleading to interpret particular cultural histories as stemming directly from organismal evolutionary processes. The specifics of cultural traditions are learned and frequently borrowed.

The fact that specific languages, for example, do not evolve in the biological sense is demonstrated by the fact that children do not prefer any particular language innately; rather, they pick up whatever languages they hear or are taught. As noted above, patterns of syntax and sentence structure common to all languages appear linked to heritable traits of brain structure and function, but such general biological features do not explain the history of change within and among particular languages. The history of particular languages (Greek, Latin, French) and their relationships are a matter of politics, geographic conquest, and demographic fates of human populations rather than of adaptive selection on variant human genes.

Problems with use of evolution as a metaphor stem from forgetting that the mechanisms of change in heritable features of organisms and the learned features of culture differ significantly. Heritable features of organisms are subject to change via natural selection and can be assessed as potential adaptations. Variable features of human culture that are learned are not influenced directly by natural selection because variation in those features is not directly heritable. Thus, variation in the cultural products of human endeavor, such as particular languages and religions, do not evolve in the biological sense. However, they do change over time in a manner analogous to biological evolution, with human innovation and the borrowing or imposition of cultural practices playing an important role. Recognizing when evolution is applied only as a metaphor helps us avoid several traps. One trap is in assuming that observed features are adaptive, simply because they exist. Another trap is assuming that observed features are biologically determined and fixed. Unfortunately, these traps are seductive in the consideration of biological as well as cultural features. However, recognizing where evolution is used as a metaphor for biological evolution should make it easier to avoid falling prey.

My objective in this chapter is to show how an evolutionary approach, even as metaphor, can provide insight into the history of cultural change, focusing on languages and religions. Language is our foremost window to the human mind, allowing us to describe our thoughts, plans, and emotions. Religions embody our ethical traditions and our views on the meaning of human lives. Understanding the history of change in the varieties of language and religion can help us understand ourselves and see the many cultural differences as variations on a theme rather than barriers between unrelated foreign traditions. I will include some branching diagrams, but they should not be mistaken for biological pedigrees. On their own, they do not explain why or how particular changes have occurred. However, the methods used to estimate the trees can effectively show historical relationships among cultural features, and this is a useful application of an evolutionary approach.

Language

Words

Etymology is the study of the history and origin of words. It uses comparative methods similar to some used in studying origins of taxa. Systematic searches for similar-sounding words with similar meanings in different languages, such as "mother" in English and "Mutter" in German, are used to infer historical relationships among words described as cognates. Just as discovery of homologous traits among organisms (traits related by common descent) is made difficult by convergent similarity, discovery of true cognates can also be complicated by convergent similarity. That is, not all word pairs from different languages having similar sounds and meanings are true cognates.

There are certain trends in the metaphorical evolution of words, however, and being aware of them can help in making a more accurate identification of cognates. These trends include a tendency for spelling to become simpler and sometimes abbreviated (e.g., "recon" from "reconnoiter" and "phone" from "telephone"). Sometimes words originate as acronyms, formed from the initial letters of words in phrases, such as "awol" (absent without official leave), "mash" (mobile army surgical hospital), and, in one more military example, "fubar" (. . . beyond all recognition).

Often words are combined in some fashion to make a new word, such as "televangelist" or "pulsar" (pulsating × neutron star). "Hippopotamus" is a Greek example, combining *hippos* for horse and *potamus* for river. The writer and mathematician Lewis Carroll coined the term "portmanteau" to describe this phenomenon. In its original usage, a portmanteau is a leather suitcase, composed of two halves joined by a hinge. It provides an example of the linguistic phenomenon it describes, being a combination from the French *porter* (to carry) and *manteau* (coat). Carroll coined many portmanteaus himself, at least one of which, "chortle" (chuckle × snort), has passed into popular usage. Many new words are derived by taking existing words and adding prefixes such as "trans-" or suffixes such as "-ness" (used to form nouns from verbs or adjectives; e.g., darkness).

In other instances, words are derived from a meaning associated with a particular place (toponyms) or person (eponyms)—for example, "Armageddon" (a final or decisive war as described in the Bible) and "sandwich" (for the handy innovation of the Earl of Sandwich). In other cases, words evolve via substitution. An example is that of the avocado, a fruit of the New World. The original Aztec name *ahucatl* is derived from their word for testicle. To the ears of the Spanish conquistadors, *ahucatl* sounded like "avocado," their term for advocate, and the fruit was introduced to Europe under that name.

SPAM illustrates several different mechanisms of recent word evolution. In 1937, the name for this canned pork product was coined as an abbreviation (SPiced hAM) by Hormel Foods. Though SPAM was eaten unceremoniously by thousands of U.S. and Russian soldiers during World War II, the term rocketed to twisted fame following the 1960s Monty Python "spam skit," in which the word "SPAM" is mentioned ninety-four times. What follows is only the beginning of the sketch.

Mr. Bun: Morning!
Waitress: Morning!
Mr. Bun: Well, what've you got?
Waitress: Well, there's egg and bacon; egg, sausage and bacon; egg and spam; egg, bacon and spam; egg, bacon, sausage

and spam; spam, bacon, sausage and spam; spam, egg, spam, spam, bacon and spam; spam, sausage, spam, spam, bacon, spam, tomato and spam; spam, spam, spam, egg and spam; *(Vikings start singing in background)* spam, spam, spam, spam, spam, spam, baked beans, spam, spam, spam and spam.

Vikings: Spam, spam, spam, spam, lovely spam, lovely spam *(drowning out all conversations, until told to shut up).*

This at least provided an initial (and for some indelible) use of the term as connoting something annoying and frequently repeated. In this context, "spam" suited the sensibilities of a particular group of early Internet users during the late 1980s known as MUDs, an acronym for multi-user dungeon group, to describe repeated, unwanted advertisement e-mails. Hormel has finally stopped complaining about the new popular definition, although they ask that their proprietary name, SPAM, be spelled in capital letters to distinguish it from the heinous Internet bane, because, in their words, they want to avoid the day "when the consuming public asks, 'Why would Hormel Foods name its product after junk e-mail?'"

Many words change meaning or are used in new ways over time. "Evolve" is derived from the Latin *evolvere,* meaning to unroll, and according to the Oxford English Dictionary (OED), its earliest written usage is in 1641. Obviously, the term was not used in the context of common ancestry for all life until much later. Of the eight variant meanings listed by the OED, it is the seventh that pertains to the subject of this book. Darwin did not use any form of the term "evolve" in his most famous book except as the final word of the last paragraph.

Although the study of etymology and biological evolution have proceeded on largely separate tracks, the parallels between evolution of words and the evolution of genes and genomes is striking. Many of the phenomena noted above, de-

Doublet puzzles

a. **APE > ARE > ERE > ERR > EAR > MAR > MAN**

b.

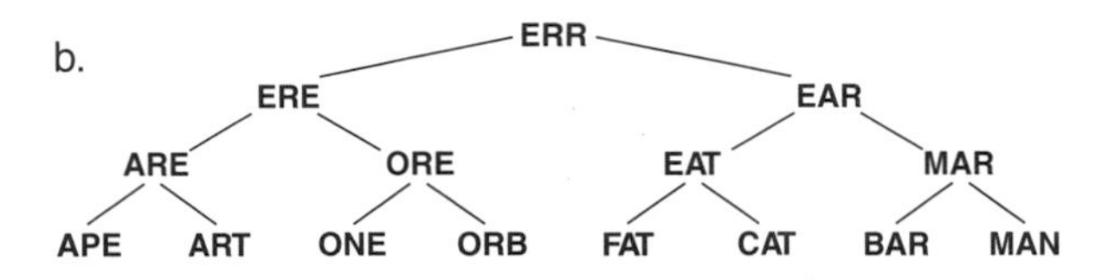

Analogous operations: phylogenetic reconstruction using DNA; search for common gene expression patterns among cells, organs, taxa; and graphing of gene networks showing protein-protein interactions.

c. **Segmentation**

IN EVERY ODE LINGER MANY
I NEVER YODEL IN GERMANY

Dreaming of apples on a wall
And dreaming often, dear,
I dreamed that if I counted all,
- How many would appear?

Analogous operations: sequence alignment with gaps; detecting gene/exon/intron boundaries, overlapping reading frames, alternative splicing sites, and predicting protein secondary structures.

d. **Nonsense** (decoupling syntax and semantics)

He thought he saw a Rattlesnake
That questioned him in Greek
He looked again, and found it was
The Middle of Next Week.
"The one thing I regret," he said,
"Is that it cannot speak!"

Analogous operations: nonsense mutations encoding an 'in frame' stopcodon mutation allowing a gene to be transcribed and translated but producing a nonfunctional, truncated polypeptide scribing the history of change for words, are found to have analogous processes at work in the evolutionary history of DNAs. These parallels are perhaps best illustrated in the works of Lewis Carroll (1832–1898), author of the fictional tales of Alice in Wonderland. Carroll, whose real name was Charles L. Dodgson, combined a serious interest in mathematics with a love of parlor games and word-play. The computational biologist David Searls, who brings a linguist's sensibility to comparative DNA sequence analyses, has pointed out the striking parallels between Carroll's word-play and the real-world permutations of genetic sequence that underlie the evolutionary history of life.[1] Carroll used doublet word-puzzles, syzygies, segmentation changes, and just plain nonsense in an unintentional preview of the letter and word permutations that modern bioinformaticists use to search for patterns of DNA sequence change and to determine their effect on meaning, in terms of protein-coding and function. Small changes in word letters and DNA sequences can both have very large effects.

Languages

Humans have been talking for more than 100,000 years. There are 5,000–6,000 different languages currently recog-

---

Figure 5.1 Parallels between word-play and bioinformatics pattern searching among genomics data sets. (a) Charles Dodgson's "doublet puzzles," in which someone is given two words of equal length and asked to find valid intervening words. (b) Doublet puzzles may also be extended to sets, in which someone is given a set of words, as in the bottom row, and asked to find the minimal tree with valid words at each interior node with one letter mutation per connection, as shown. (c) Changing the segmentation (word boundaries) for contiguous sets of letters can drastically alter meanings. Dodgson posed the riddle, "Dreaming of apples . . ." as shown. The answer for this riddle can be found by introducing a gap, changing "often" to "of ten." (d) Nonsense verse by Dodgson.

nized, though only 4,000 are still spoken, and these are placed into about 200 different language families.[2] The relationship between the history of human populations and the history of languages was pointed out by Darwin in *On the Origin of Species:* "If we possessed a perfect pedigree of mankind, a genealogical arrangement of the races of man would afford the best classification of the various languages now spoken throughout the world; and if all extinct languages, and all intermediate and slowly changing dialects, were to be included, such an arrangement would be the only possible one."

Linguists generally recognize two sources for similarity among languages; "genetic similarity," which is another metaphor, and "areal diffusion." So-called genetic similarity is modeled using branching diagrams (trees) in which a single parental language gives rise to daughter languages, analogous to phylogenetic trees for species in which one lineage becomes two. The second source, areal diffusion, contributes when there is borrowing of language elements among geographical neighbors. This might be shown as horizontal branches in the tree diagrams, though it is done infrequently. Borrowing of words from a different language is analogous to horizontal gene transfer between species as accomplished by a bacterial plasmid or virus, or hybridization between species if the mixing of two languages is extensive. In the field of phylogenetics, horizontal transfer events confound phylogenetic analyses if vertical inheritance of traits is assumed. The problem can be rectified by using techniques designed to identify traits that have been horizontally transferred, if biological, or borrowed, if linguistic, and then analyzing them separately. Increasingly, as geographical distances are overcome by frequent travel and electronic communications, linguistic elements are transferred among nonneighbors too. This is prevalent where there is social pressure to conform to a dominant language and where travel and cultural interchange of all kinds is common.

Table 5.1 Words used in nine different languages to communicate the meaning of: I, bird, fish, blood, and hand

English	French	German	Italian	Spanish	Dutch	Esperanto	Swedish	Latin
I	je	ich	io	yo	ik	mi	jag	ego
bird	oiseau	Vogel	uccello	pájaro	vogel	birdo	fagel	avis
fish	poisson	Fisch	pesce	pescado	vis	fiso	fisk	piscis
blood	sang	Blut	sangue	sangre	bloed	sango	blod	sanguis
hand	main	Hand	mano	mano	hand	mano	hand	manus

The methods for assessing relationships among languages and species of organisms were initially developed in isolation from each other; however, that has changed in recent times. Morris Swadesh helped to pioneer comparative studies of language history during the early 1950s. He compiled a list of a hundred or so common words essential to most languages and used them to calculate the proportion of cognate words shared by language pairs as a measure of their relatedness.[3] As more instances of cognates are identified among pairs of languages, the case for their "genetic" relationship is strengthened, as such increasing numbers of cognates are unlikely to happen by chance.

This much was straightforward and well received. Swadesh also estimated rates of change and timing of divergences among languages, known as glottochronology. For this, he plotted the proportions of shared cognates for language pairs against an estimated time since their divergence. If rates of divergence were constant over time, the resulting plot would show a straight line. Swadesh advocated the notion of rate constancy with a vocabulary decay rate of about 14 percent over 1,000 years. Today, no linguist would assume constant rates of language change over time, just as no biologist would assume constant rates of DNA sequence change. The Swadesh word list remains useful for initial comparisons among languages, however. Because of a tendency for those basic words to change relatively infrequently, the list still provides a basis for measures of similarity among distantly related languages. Further, he helped introduce quantitative statistical approaches that continue to be used and improved upon in analyses of larger and more refined data sets.

More recently, the statistical methods of maximum-likelihood and Bayesian analysis as developed for estimation of organismal evolution have been applied to analogous issues of relationships among languages.[4] These methods use explicit models of character evolution in estimating historical relationships. The resulting trees showing historical relationships among languages are based on comparisons of words and searches for cognates, but elements of syntax and grammar might also be analyzed separately or in combination with words. Traditionally, language family trees have sought to cover linguistic change going back 8,000 years or so.[5] However, analyses using evolutionary models focusing on the most slowly changing features suggest that historical relationships can be revealed going back much further in time, possibly 20,000 years ago or more. The table outlines the analogous mechanisms and features involved in change over time in organisms, languages, and religions.

It is worth noting an operational difference between phylogenetic analyses for species versus languages. Phylogeny for organisms is based on comparison of homologous traits, sharing common ancestry. Phylogeny for languages is based on comparisons of words with similar meaning, whether or not they are deemed to be cognates sharing the same linguistic root. For example, the words used in different languages to name the forelimb ("arm" in English) are compared. So, organismal phylogeny is based on comparison of traits shared among taxa owing to common ancestry, but linguistic phylogeny is not, being based on comparison of words to an object, definition, or con-

Table 5.2 Analogous features useful in reconstructing evolutionary history for organisms and cultural history for languages and religions

Feature	Biological organisms	Languages	Religions
Units	nucleotides, codons, amino acids, genes, morphological traits, individuals	words, pronunciation, meaning, grammar/ syntax	beliefs, traditional practices, religious documents, historical accounts
Mechanisms of change	biological evolution: genetic mutation, speciation (reproductive isolation for vertebrates)	cultural evolution: human innovation or mistakes, adoption, imposition	cultural evolution: human innovation or mistakes, adoption, imposition
Primary transmission mode	inheritance of DNA from parent(s) to progeny (vertical transmission)	teaching, learning, imitation	teaching, learning, imitation
Secondary transmission mode	horizontal transmission of DNA (e.g., hybridization of species, lateral gene transfer)	adoption, imposition	adoption, imposition
Change mediated by	natural selection, chance	human intention, chance	human intention, chance

Source: Expanded from M. Pagel, "Maximum-likelihood models for glottochronology and for reconstructing linguistic phylogenies," in C. Renfrew, A. McMahon, and L. Trask, eds., *Time Depth in Historical Linguistics,* vol. 1 (Oxford: The McDonald Inst. for Archaeological Research, Oxbow Books, 2000), pp. 189–222.

cept. This is not to suggest that language phylogeny is not useful for historical reconstruction. Rather, it is to illustrate the distinction between biological evolution and an analogy and avoid the trap of assuming a common mechanism of change.

What has been learned in applying an evolutionary approach to the history of languages? There is now a well-supported hypothesis for the origin of and historical relationships among the Indo-European languages spoken from Ireland east to the Indian subcontinent. Although many of these relationships have been supported by earlier studies, others remained unresolved. Because of their use of discrete language traits rather than summary distance measures, their improved handling of the highly variable rates of lexical change across languages, and an ability to distinguish and remove elements that are borrowed rather than endemic, the recent analyses can go further back in time with greater reliability.

The analysis in the accompanying figure can be applied to a longstanding question of Indo-European language history. Similarities between Sanskrit, an ancient language of India, and Greek and Latin have been known to generations of schol-

Figure 5.2 Tree showing relatedness of Indo-European languages, which uses the same analytical approach (maximum-likelihood analysis) used to infer phylogenetic relationships of organisms. Age estimates in years before the present are shown for some divergences.

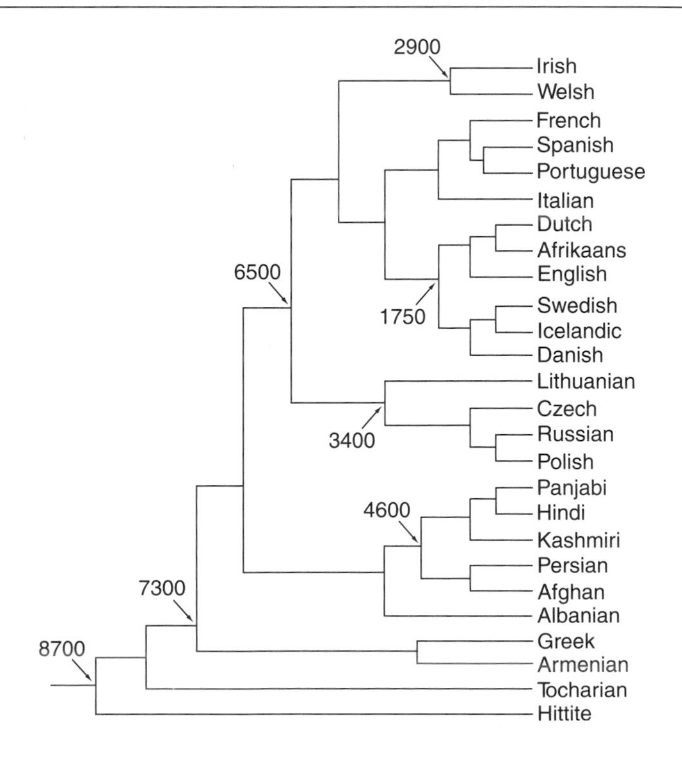

ars, but poorly understood. The "Kurgan expansion" hypothesis, based on archaeological evidence, supposes that the progenitor of the Indo-European group of languages was brought to Europe and the Near East by nomadic peoples of the steppes north of the Black Sea, Kurgan horsemen, about 6,000 years ago. An alternative Anatolian hypothesis supposes that the Indo-European language group diversified along with the spread of agriculture from Anatolia, spanning much of modern Turkey, 8,000–9,500 years ago. Support for one or the other hypothesis depends critically on the estimated ages for the earliest divergences within the group. Using known dates of historical events, written documents, and inscribed artifacts as calibration points, recent analyses indicate early divergences between 9,000 and 7,000 years ago, supporting the Anatolian hypothesis. The date estimates do not assume constancy in rates of language change, and can be retested as more language data sets and calibration dates are compiled.[6]

Figure 5.3 Trees showing horizontal gene transfer for organisms and word-borrowing for languages. (a) Tree showing phylogeny for the three primary domains of life. Eukarya and Archaea are sister taxa and share genes for replication, transcription, and translation of DNA. Eukarya have also picked up genes involved in glycolysis and other processes via endosymbiosis and horizontal gene transfer from Bacteria, shown with an arrow. (b) Tree showing relationships among three languages. English and German are most closely related. Extensive borrowing by English speakers from the language lineage leading to French is shown with an arrow.

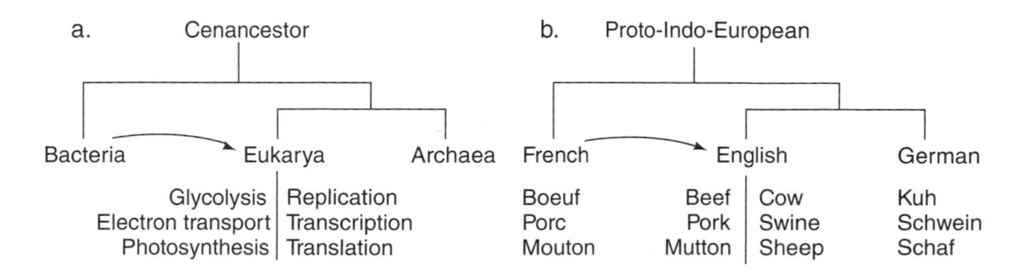

The borrowing or adoption of words among different languages is common, and those borrowings can also be recognized with detailed phylogenetic analyses. Borrowing of words among languages is analogous to lateral transfer of genes among species and their genomes.

Languages encapsulate a great deal of the culture of societies, including some indication of their values, beliefs, and ways of classifying and viewing the world around them. Improving language phylogenies will improve our grasp of the pattern and timing of language divergences. As more languages are added to these phylogenies, it will be possible to learn more about the ways human minds work in structuring languages and modifying them. Understanding the historical relationships among languages will help us understand more about both shared values and culture across human populations and the history of individual human populations.

Books

One way to gauge the timing and cultural impact of language change is to examine the history of translations for particular written works. The history of Hebrew scriptures is interesting in this regard, being central to several related cultures. An overview of one of the cultural paths emanating from Hebrew scriptures will also show how the metaphorical evolution of a written work, changing in the course of translation as different languages fall in and out of favor, differs in its mechanisms from the biological evolution of populations of organisms. The changes in this influential cultural document stem from politics rather than variation in heritable biological traits.

As noted earlier, archaeology and historical scholarship suggests that Hebrew scriptures, based on older oral traditions, were written down between 1000 BCE and 350 BCE. The oldest set of writings, traditionally ascribed to Moses, are known as the Pentateuch, a Greek term for "five books" (Genesis, Exodus, Leviticus, Numbers, Deuteronomy). There appear to have been multiple authors for these works, which were originally written in Hebrew with the exception of a few brief sections written in its sister language, Aramaic. The full set of Hebrew scriptures, comprising thirty-nine books in total, provide a rich mixture of stories, poetry, laws, parables, songs for worship, descriptions of events, and predictions about the future. They include numerous internal contradictions, miraculous happenings, and an abundance of murder, incest, deceit, tragedy, and passion. They also include the oldest documented ethical code for human behavior together with a covenant between a particular group of people and a single divine power as (among other things) an inducement to comply. Principles of ideal behavior, actual behavior, and a moral authority are combined in one chaotic package with many loose ends. Hebrew was the common language of the Jewish people from at least the late second millennium BCE. However, Hebrew was replaced by Aramaic as the spoken language among Jews following their defeat by the Babylonians in 586 BCE. Hebrew survived as the language of Jewish liturgy and literature, though little spoken, until its revival in the late 1800s and its designation as the language of Israel in 1948. Prior to 1948, there had been extensive lobbying by speakers of the many living languages of the refugees and endemic Jews, from French to Farsi, in British Palestine as a way to designate one particular culture as primary. The decision to revive Hebrew was deemed most even-handed, as that was the only spoken language that no one knew.

The New Testament, ultimately presented in twenty-seven books, tells about the life and teachings of Jesus, who initiated a variant cultural tradition within the broad outlines of that documented in Hebrew scriptures. These works were first written in Greek, the dominant scholarly language of the time, by multiple authors during the span of several decades during the first and second centuries.

Word choice in translations matters, and it would be difficult to find one with more historical impact than that of Isaiah 7:14. In this passage from the Old Testament, the prophet Isaiah says to King Ahaz of Judah, "Behold, *ha'almah* shall conceive, and bear a son, and shall call his name Immanuel." In the New Testament Gospel of Matthew (1:22–23, KJV) the prophecy is restated as, "Behold, a virgin shall be with child, and shall bring forth a son, and they shall call his name Emmanuel, which being interpreted is, 'God with us.'" The Hebrew *almah*, when translated from the Old Testament to the New Testament, became the Greek *parthenos,* meaning virgin. *Almah* is most accurately translated to mean "young woman," with no implication about virginity, *almah* being the feminine form of the word for young male (and *ha* meaning "the"). Taking this passage from Matthew literally, it can be read as the prophet Isaiah describing Jesus' miraculous birth to a virgin (some 700 years in the future). The New Testament's author may have been following the Septuagint, the translation of the Old Testament from Hebrew to Greek used by Greek-speaking Jews at the time, which translated the Hebrew word *almah* as "parthenos." Thus, at a very early point in the development of Christianity, we see a change in a word having significant effect, in canonizing the prophecy of a virgin birth.

The spread of Christianity and the cultural influence of the Old and New Testaments fed each other. Ultimately, the Bible was translated into all the Indo-European languages, though amid heated controversy and social turmoil over cultural ownership. Should only clergy be allowed to read the works? Should they be translated into crude vernacular languages? Which of the many writings should be included and which should be dropped? Which sources should be used in translation? Which shades of meaning should be emphasized, and which passages should be changed entirely to promote political agendas?[7]

Each first translation into a new language was accompanied by controversy of some kind, and part of a larger process of cultural adoption and change. The Septuagint is the earliest translation of the Hebrew scriptures into Greek, translated by Hellenistic Jews during 275–100 BCE in Alexandria. The name "Septuagint" derives from the traditional view that it was the work of seventy different translators. Its later popularity and occasional revision by Christians was off-putting for Jews, who stopped using it after about 70 BCE, preferring the original Hebrew or alternative Greek translations. By 100 CE the Old Testament had been translated into Syriac, the language of people living around Edessa in modern Turkey, and missionaries from the Syriac Christian church carried it in their travels as far as India and China. During the fourth century the entire Bible was translated into Coptic, a form of the ancient Egyptian language, to meet the needs of Christian converts in northern Egypt. Around this time, Ethiopian, Armenian, Gothic, Slavonic, Georgian, and Arabic translations were being made.

In 405 CE, the Roman scholar Eusebius Hieronymus, known to many as "Jerome," completed a translation of the Bible into vernacular Latin. Jerome had spent twenty-one years preparing this work, using the earliest available Greek and Hebrew sources as well as the few previous Latin works. His early translations were resented by clergymen in Rome, so he moved to Palestine where he founded a monastery and completed his work, which he called the Vulgate, to describe its use of Vulgar Latin, the common spoken language of the western Roman Empire. The Vulgate eventually became the sanctioned Bible for the Roman Catholic church and the common source for many translations to other languages.

The first complete English translation of the Bible was

Figure 5.4 Tree diagram outlining the history of Bible translations focusing on English translations.

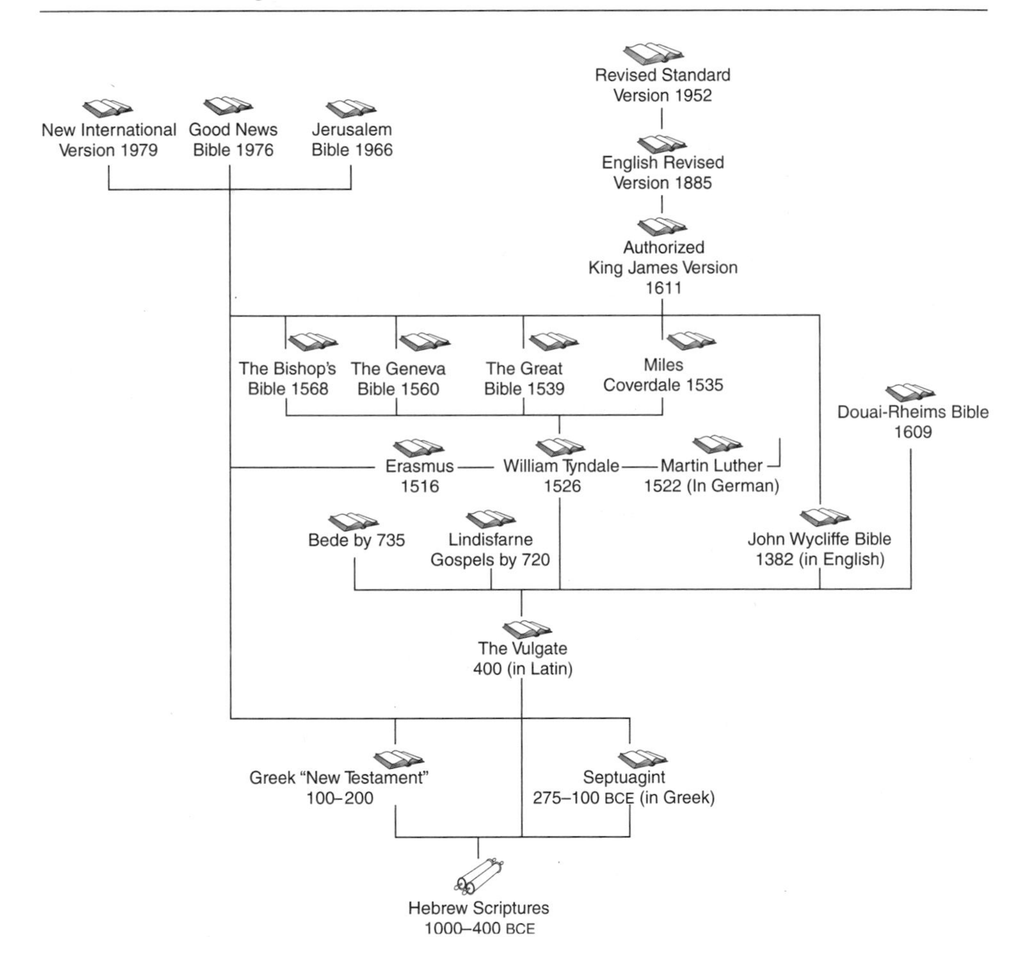

made from the Vulgate by John Wycliffe in 1382. Wycliffe was an outspoken critic of many church practices, as well as its abuses of power, and sought to make the Bible accessible to regular people. As reward for his efforts, he was forced out of his teaching position at the University of Oxford, and forty-four years after his death, his bones were dug up and burned on orders from the Pope.

In the fourteenth century the new culture of the Italian Renaissance swept through much of western Europe. Literacy became a measure of personal accomplishment and a key to fulfillment, fueling demand for books. Hand-copying books delayed the dissemination of ideas and entertainment, as well as driving up costs. Inventors sought new technologies for book production, and in about 1450 German metalworker Johannes Gutenberg succeeded and won the race. He invented a moveable metal type and combined this with other innovations (including making ink from candle soot mixed with varnish and using paper instead of vellum) to develop a process to produce books relatively quickly and cheaply. When Gutenberg looked for a challenging test for his new printing press, the Latin Bible was a natural. It was lengthy (requiring 1,768 pages) and popular. New churches were abundant, their congregations were growing, pilgrimages to the Holy Land were popular, and general interest in religion was reaching new heights. It was a good business decision, as the cost of handwritten copies was prohibitive for all but a few. Gutenberg's print run, completed in 1456, yielded just 185 copies, of which forty or so still exist. Gutenberg did not become wealthy, having lost a legal battle with his partner over repayment of loans. But his printing technology, which was immediately imitated and improved by others, quickened the pace of cultural transformation like nothing before.

John Wycliffe's reformist passion was carried on by Martin Luther in Germany. In 1517, Luther famously posted his 95 objections to the sale of official church pardons ("indulgences") for sins that would otherwise require more time in purgatory. Five years later, in 1522, Luther completed and published his translation of the New Testament into German. Not only did this put a vernacular Bible into the hands of nonclergy, it figured prominently in establishing German as the single unifying language in Germany. This was at a time when most scholarly activity was conducted in Latin, and many felt nascent German to be incapable of a worthy biblical rendering. Luther published about 100,000 copies of his German translation, seeding additional translations. Among the changes he introduced was a reordering of the New Testament books, placing Hebrews, James, Jude, and Revelation at the end, unnumbered, indicating doubts about their authenticity.

The same desire to free the Bible and its dispensations from control by clergymen motivated William Tyndale to translate and publish an English Bible. This coincided with an increase in education, growing nationalism, and a desire for cultural distinctiveness among the English-speaking populace, despite the preferences of English nobility for French and that of entrenched academics and clergy for Latin. Tyndale might have pursued his translation at Oxford where he had been a student; however, disdain there for English writing and a philosophical rather than historical and direct approach to theology discouraged him. Instead, he sought to work elsewhere and eventually published his work in Germany. Tyndale's first attempt to publish his New Testament translation in 1525 on German printing presses was thwarted by Johannes Cochlaeus, a Catholic opponent of Luther, following a slip of the tongue by an inebriated printer's assistant. Tyndale succeeded about a year later, working under greater security. His New Testament English translation was based on the 1516 Latin translation by Erasmus, original Greek texts, and the work of Martin Luther, among others. English church officials condemned his work as soon as it arrived on English shores and called for his arrest on charges of heresy. He continued to work in hiding, translating and publishing an English translation of the Five Books of Moses in 1530, but was eventually put on trial and strangled before being burned at the stake in 1536.

By 1600, there were numerous English versions of the Bible. In addition to William Tyndale's, there was Miles Coverdale's Bible and the Great Bible. Also known as Cromwell's Bible, it was the first authorized for public use in England. The Geneva Bible was favored by Shakespeare and considered the most popular at the time. These existed within a context of religious tensions in England, not only between Catholics and reform-minded Protestants, but also between Anglicans, who favored a strong role for monarchy in the church, and Puritans, who did not. Puritans favored the Geneva Bible, which suited their views best.[8] For example, annotations in the book of Daniel directly challenge the notion of divine rights of kings. In the biblical narrative, Daniel has been thrown into a den of lions for disobeying the king's orders, and the text with annotations (denoted "h" and "i") from the Geneva Bible reads as follows.

> Daniel 6:22: My God hath sent his angel, and hath shut the lions' mouths, that they have not hurt me: forasmuch as before him {h} innocency was found in me; and also before thee, O king, have I done {i} no hurt.
>
> {*annotation:* h} My just cause and uprightness in this thing in which I was charged, is approved by God.
>
> {*annotation:* i} For he disobeyed the king's wicked commandment in order to obey God, and so he did no injury to the king, who ought to command nothing by which God would be dishonoured.

Advocating that the orders of kings, annotated elsewhere as "tyrants," should be secondary to the will of God resonated well with the Puritans, who dominated English Parliament, but not with King James I or the Anglicans.

In 1604, King James I convened a meeting of church officials, the Hampton Court Conference, where it was resolved that, "A translation be made of the whole Bible, as consonant as can be to the original Hebrew and Greek; and this to be set out and printed, without any marginal notes, and only to be used in all churches of England in time of divine service." Ironically, a Puritan delegation had originally lobbied James I for church reform precipitating the conference, and in the course of the conference, one of their delegates proposed that a new translation of the Bible be authorized and prepared, apparently hoping that it would be similar to the Geneva Bible.

The King James Bible, without margin notes, was eventually completed in 1611. The King James Bible has undergone many revisions and remains popular in the English-speaking world, having attained the status of a classic. The King James Bible is often ranked with Shakespeare in its influence on English language and literature, and in this way some of the Semitic idioms present in Hebrew scriptures have penetrated to English. Some examples are: "pride goes before a fall," "to stand in awe," "a man after his own heart," "like a lamb to the slaughter," "to go from strength to strength," and "sour grapes."

Translations can take on a life of their own as subsequent generations imbue them with personal meaning. Consider two versions of one of the best-known Old Testament passages, the 23rd Psalm. Some might prefer the familiarity of the King James version, first published in 1611. Others might prefer the accuracy of the New Jewish Publication Society (New JPS) translation published in 1985, made with the hindsight gleaned from four additional centuries of academic scholarship in translating traditional Hebrew text into the idiom of modern English.

King James Version (1611):
THE LORD is my shepherd; I shall not want.
He maketh me to lie down in green pastures; he leadeth me beside the still waters.
He restoreth my soul: he leadeth me in the paths of righteousness for his name's sake.

Yea, though I walk through the valley of the shadow of death, I will fear no evil: for thou art with me; thy rod and thy staff they comfort me.

New JPS Version (1985):
The LORD is my shepherd:
I lack nothing.
He makes me lie down in green pastures;
He leads me to water in places of repose;
He renews my life;
He guides me in right paths
as befits His name.
Thou I walk through a valley of deepest darkness,
I fear no harm, for You are with me;
Your rod and Your staff—they comfort me.

Consider too the diverse changes in English style, spelling, and letter use (orthography) that both of the above versions present relative to the earliest English translation, found in the Wycliffe Bible of 1384.

Wycliffe (1384):
The Lord gouerneth me, and no thing to me shal lacke; in the place of leswe where he me ful sette. Ouer watir of fulfilling he nurshide me; my soule he conuertide. He broghte doun me upon the sties of rightwisnesse; for his name. For whi and if I shal go in the myddel of the shadewe of deth; I shal not dreden euelis, for thou art with me. Thi gherde and thy staf; tho han confortid me.

The collection of ancient parables and narratives comprising the Bible have changed over time. Perhaps the largest changes occurred without being recorded during the millennia when all stories and knowledge were transmitted orally. As the documents changed from impressions left in baked clay tablets, to hand-copying with ink (made from plant tannins or the secretions of squid) on animal skins, to the first printing presses with interchangeable metal letters, the languages used and developed have varied dramatically, as has the mindset and worldview of the scribes and translators. Many changes in phrasing, style, and meaning have resulted from frequent filtering through different languages and across cultures, which are also changing over time. Our understanding of ancient language styles, idiom, and vocabulary has also changed as new linguistic and archaeological evidence came to light. Some changes in biblical texts have been intentional, resulting from subjective choice among alternative word meanings or choice about inclusion or exclusion of sections, and the use of annotations to emphasize particular interpretations. Other changes originated as errors in copying.

There are analogies with biological evolution of particular DNA sequences over time, which also suffer copying errors, loss or addition of pieces, and even occasional changing in the reading frame (start and end point for codons or "words"). More impressive than the change in the Bible, however, is the overall fidelity to the original human documents first scratched out starting over 3,000 years ago. This is due largely to the efforts of scholars and historians, sometimes at odds with orthodox clergy, in seeking accessibility and accuracy in translations. The variation in bibles and their interpretations provide a window on human culture as old as recorded history, as well as a partial view of some of the religious cultural conflicts that continue to animate human history.

Religion

For many, organized religions are synonymous with unchanging beliefs and fundamental, supernatural truths. In that context, the topic of the evolution of religions must seem ill conceived. But of course religions, as cultures, do evolve in the metaphorical sense. The previous section on biblical translations and political change attests to that. In this chapter, I am reviewing aspects of cultural change, noting occasional parallels with organismal evolution, and identifying historical links that can be used to assess relatedness among cultural features. Having started with words, language, and the Bible, we can turn to the historical relationships that are explained by the evolution of religions. There are several historical patterns or features that are analogous in biological and cultural evolution to be considered, and these include phylogeny, intermediate forms, convergences, and reticulate evolution or borrowings. Beyond analogy, there are situations in which molecular evolutionary studies of human populations are able to address longstanding hypotheses of cultural history and these will also be examined.

For most vertebrate animals, the primary mechanism for the origin of new species begins with differentiation among populations within a single species that occupy different geographic regions. Reproductive mixing among geographic populations may be reduced as a result of new barriers to dispersal of individuals and differences in selective pressures, favoring alternative organismal forms in the different geographic regions. Differentiation in heritable traits is further increased as reproductive mixing is reduced. Eventually, reproductive isolation becomes a matter of genetic, morphological, and behavioral incompatibility. For example, this process of speciation via geographic and habitat separation underlies the rapid diversification of honeycreepers, a diverse group of closely related songbird species in Hawaii. The Hawaiian Islands have arisen sequentially due to volcanic activity over the past 5 to 6 million years. Finch-like songbirds from the mainland initially colonized the oldest island, and their descendents have subsequently colonized (and recolonized) the newer islands in the archipelago as they arose. The surviving fifty-seven or so species in the family show a variety of distinctive adaptive features, including long curved bills in some species well matched to the long tubular flowers from which they extract nectar. Phylogenetic analyses for groups like the honeycreepers can reveal the historical pattern of dispersal giving rise to the different species.

If we apply a naturalist's perspective to the history of religious cultures, we can find analogous patterns of shared origins followed by a series of changes, some of which demarcate related but distinctive religious groups. Religions originate in the course of human history, and following initial periods of identity formation, they may experience varying degrees of dissent, reform, factionalism, and occasional differentiation. Often geography is involved, as local differences in both secular and competing religious cultures influence the relative embrace and spread of novel cultural practices.

History of Abrahamic Religions

Hebrew scriptures tell the story of the covenant between God and Abraham and his descendents. Judaism, Christianity, and Islam all claim Abraham as their religious and cultural patriarch, and are reasonably viewed as a family of related religious cultures. A cultural genealogy, analogous to organismal phylogeny, can be drawn depicting historical relationships among the cultures, and various shared and unique features can be identified.[9] According to the scriptural narrative, Abraham had eight sons. Jews trace their history through Abraham's son Isaac in particular. Isaac's son Jacob had twelve sons who gave rise to the twelve tribes of Israel. Males of the tribe of Levi constituted a hereditary priesthood, and were scattered among the other tribes.

Although there is no definitive proof of Abraham's existence, historians consider Abraham and his extended family to

Figure 5.5 Tree of Abrahamic religions outlining the evolution of some traditions as characters. Character origins are denoted with black circles, and character losses, or replacements, are denoted with open circles. For example, circumcision is shown as a cultural practice that arose sometime before the Abrahamic religions, and is shared by other, non-Abrahamic, traditions. Monotheism remains as a core concept for all, although the dietary laws, Saturday sabbath, and maintenance of Jerusalem as the primary holy site have been dropped, variously, within parts of the Christian or Islamic traditions. Correspondingly, recognition of new prophets, Jesus and Muhammad, as founders help define Christianity and Islam, respectively.

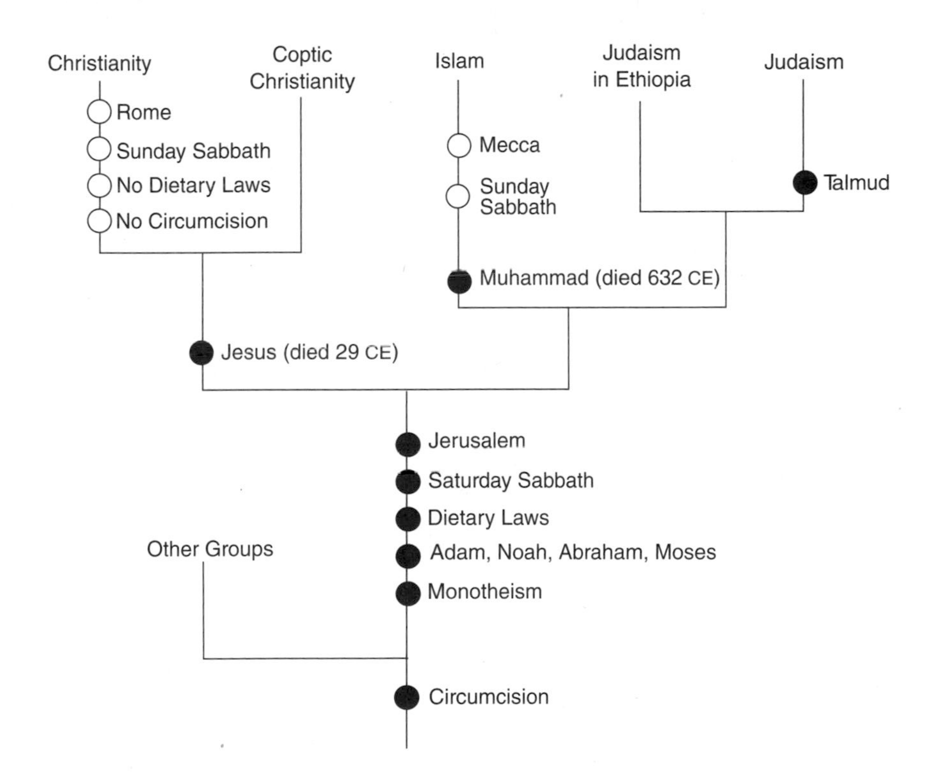

Figure 5.6 Tree showing historical relatedness for a variety of Jewish groups.

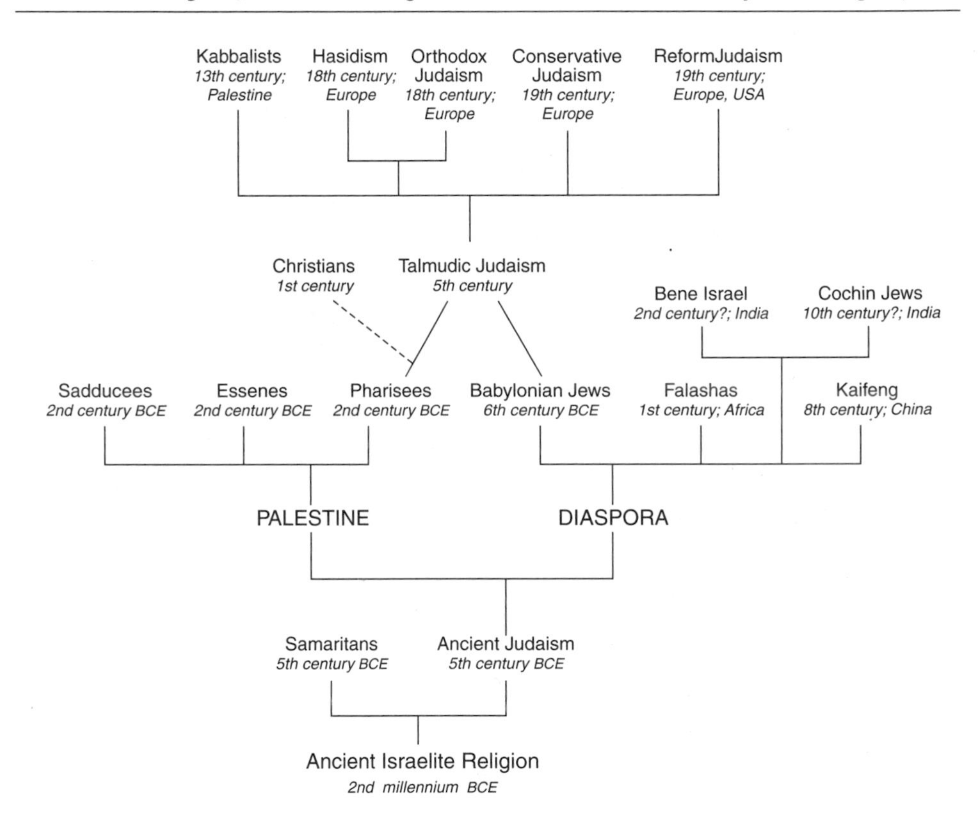

have been real people, as many of the places and events described in Hebrew scriptures have been corroborated by archaeological findings, and these allow estimation of an approximate timeline. Abraham is thought to have lived initially in and around the ancient Sumerian city of Ur, in modern Iraq, about 2000 BCE, moving regularly to meet the needs of his livestock. Abraham led his people to the lands said to be promised to them by God (Canaan) between today's Jordan River and the Mediterranean Sea. Tradition holds that after Abra-

Figure 5.7 Tree showing historical relatedness for a variety of Christian groups.

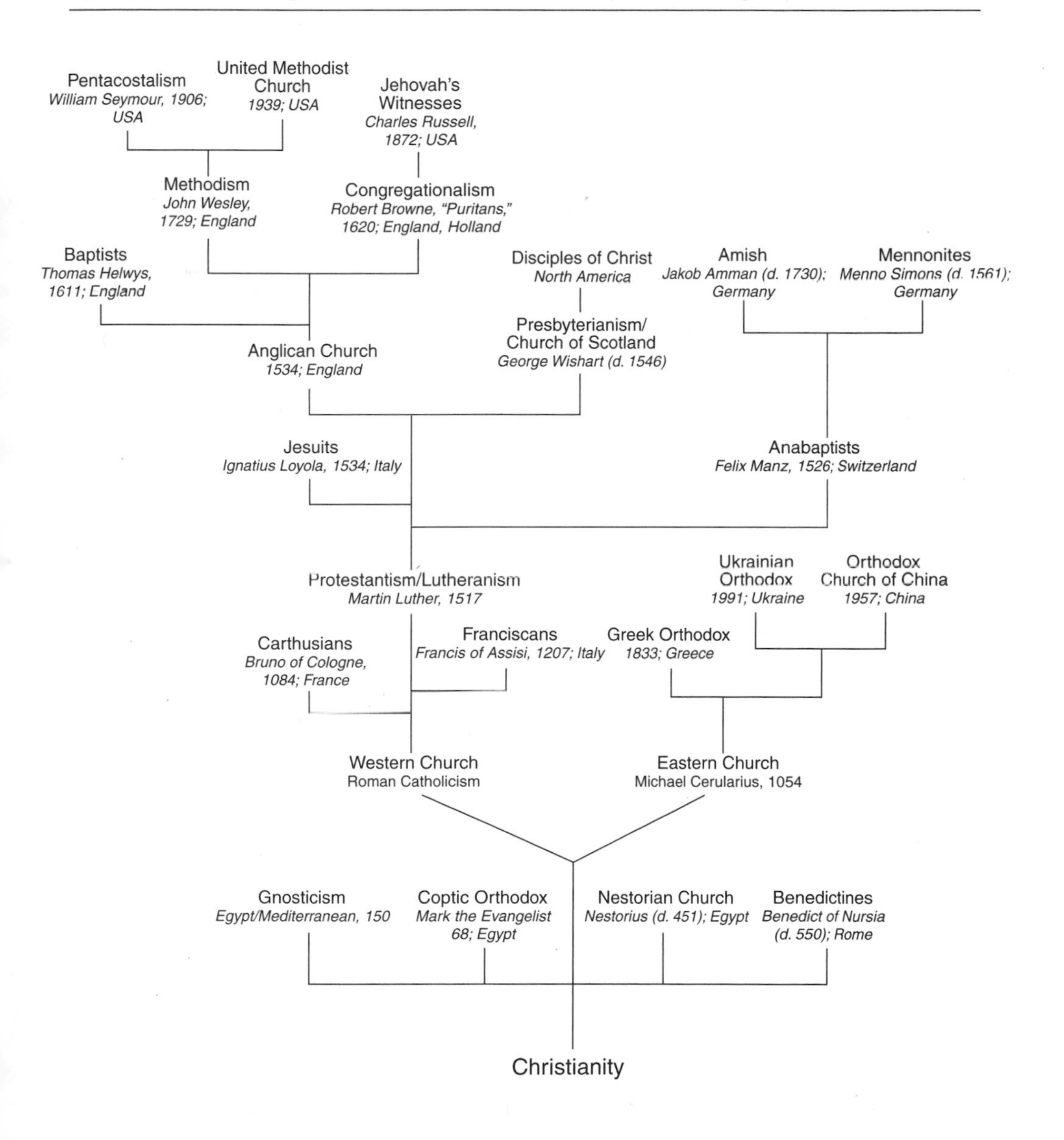

Figure 5.8 Tree showing historical relatedness for a variety of Islamic groups.

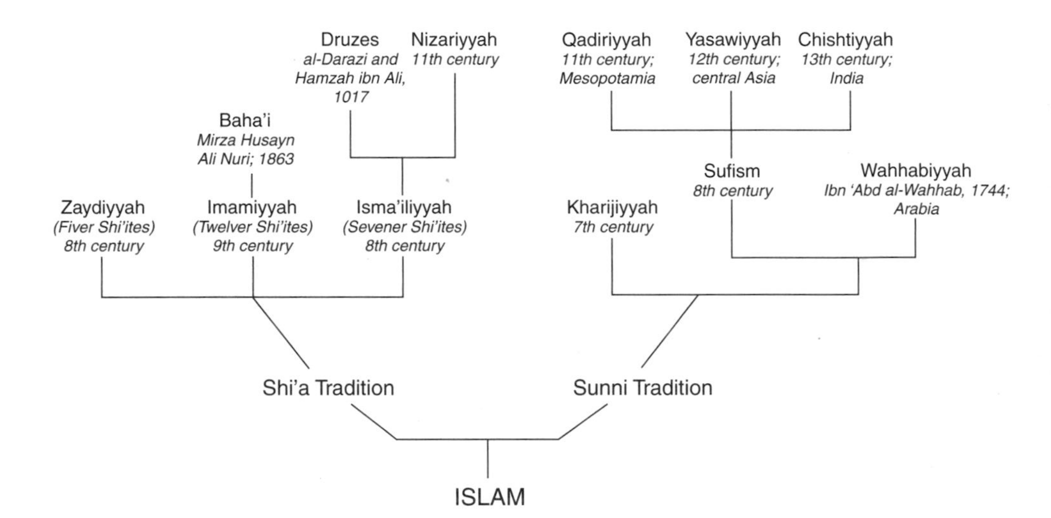

ham's death, famine forced Jacob and his family to emigrate to Egypt. Following eventual exodus from Egypt around 1275 BCE, many Jews returned to Canaan, comprising a single kingdom, until civil war divided them in about 922 BCE. The northern and southern political kingdoms of Israel were eventually brought down by the invasions of two Assyrian kings, and their tribes were exiled.

The ancient Israelites, descendents of Abraham, Isaac, and Jacob, began recording their oral traditions in Hebrew documents in about 1000 BCE. Hebrew scriptures helped define the Israelites and the early history of Jewish religious culture, with belief in a single all-powerful deity and a set of moral and religious instructions whose core, the Ten Commandments, were claimed to have been revealed to Moses by God on Mt. Sinai. There is a long history of cultural change (and remarkable fidelity too) in the lineage beginning with the ancient Israelite religion.

Christianity began as an offshoot of Jewish religious culture. The followers of Jesus, all Jews, saw him as the messiah whose coming was foretold in Hebrew scriptures. Their writings about Jesus' life and teachings during 100–200 CE provided the basis for the New Testament and a new, related but distinctive, religious cultural lineage. Christianity shared belief in the God of Abraham, the prophets of scriptures, and the received nature of the Ten Commandments, and added belief in Jesus as the messiah. As Christianity spread through Europe and western Asia, a variety of sects arose and differentiated. Some of the features of Jewish religious culture, such as the dietary laws, were dropped and cultural habits of local peoples adopted, making it easier to find converts.

Muslims trace their religious cultural history to Abraham through his son Ishmael. Traditionally, Ishmael is considered the patriarch of the Arab people. His descendents are believed to include Muhammad, the prophet and founder of Islam. Muhammad was born in about 570 CE, and the Qu'ran, the sacred document of Islam, records his divine revelations.

Within Islam, the primary divergence between the Shi'a and Sunni groups began as a dispute over authority to rule the Muslim community. The Shi'a believe that the prophet Muhammad's son-in-law, Ali, and his descendents are divinely authorized to rule, whereas the majority Sunni group believe a community consensus is required. Though Islam has just two primary subgroups, Shi'a and Sunni, there is another distinctive religious culture, Baha'i, with its roots in Shi'a Islam. The Baha'i religion originated in Iran in 1844. The founder, a Persian Shi'ite Muslim, Mirza Husayn Ali Nuri (1817–1892), declared himself to be the long-awaited twelfth *imam* and messiah, and assumed the title of *bab,* meaning "gate," by which

Muslim believers may gain access to the true faith. In addition to finding its roots in Shi'a Islam, Baha'i accepts many prophets to be "manifestations of God," including Abraham, Moses, Zoroaster, Krishna, Buddha, Jesus, Muhammad, and its own, Baha'u'llah, another name for its founder and prophet. Baha'i is a progressive faith recognizing the validity of all religions, advocating one world government, denouncing prejudices by race, sex, or religion, and attempting an integration of scientific and religious ideas, with scientific findings given primacy over religious beliefs. Though they have suffered extreme persecution, Baha'i communities are found in nearly every country, and they represent the single largest minority religious group in Iran. Baha'i history originates in Islam with broad cultural admixture.

Early Forms

Early evolutionary features of organisms, if maintained with little or no subsequent change, can provide useful evidence for stages in the evolutionary history for species. For organisms, an abundance of intermediate fossil forms have been discovered. Two of the best-known examples are *Archaeopteryx,* a small dinosaur with feathers, teeth, and tail vertebrae, showing it to be part of a larger lineage that gave rise to birds, and *Basilosaurus,* a whale with small but still functional legs. Many surviving forms show transitional features as well. For example, all species of amphibians maintain transitional features, given that they must lay their eggs in water even though they are capable of breathing air and dispersing across terrestrial habitats. The ages of fossils of *Archaeopteryx* and *Basilosaurus* are consistent with their transitional place in evolutionary history, with the relatives and descendents of *Archaeopteryx, Basilosaurus,* and early amphibians giving rise to birds, fully aquatic whales, and reptiles, respectively.

Early forms of religious cultures, still practiced by some, can also be identified. A small surviving group of Samaritans practice an early fundamentalist form of Judaism. They are monotheistic, and they claim the Torah written in Samaritan script and two additional fourth-century works unique to the Samaritans as their sacred documents. They also practice occasional animal sacrifice. They do not recognize more recent developments within Judaism, including the writings on religious practice and law collected in the Talmud, which was compiled between 300 and 600 CE. The Samaritans broke with the Jewish majority to build a temple on Mt. Gerizim near Nablus in the West Bank. Today about half of the estimated 550 remaining Samaritans live near Mt. Gerizim, with the others living near Tel Aviv.

The Jews from Ethiopia, known as Falashas or Beta Israel (House of Israel), also practice a pre-talmudic Judaism. They read Torah and the rest of the Hebrew scriptures in Ge'ez, an ancient Semitic language of Ethiopia, and are strict in their observance of the sabbath on Saturday and the biblical dietary laws. They celebrate the main biblical festivals, including the exodus from slavery in Egypt (Passover), and commemorate, by fasting, the destruction of the First Temple, which occurred in 586 BCE. According to legend, Beta Israel are descendents of the followers of Menelek, the son of King Solomon and the Queen of Sheba, who traveled from Jerusalem to Ethiopia around 1000 BCE. A modern emperor of Ethiopia, Haile Selassie, who ruled from 1930 to 1973, also claimed this line of descent. Other legends claim the Ethiopian Jews to be descended from Israelites that stayed on in Africa after the Egyptian exodus, or from tribes that drifted from Jerusalem after destruction of the First Temple in 586 BCE. Though their origins as a group are unknown, it is clear the Beta Israel practice an early form of Judaism mixed with several Christian or Muslim traditions. Many scholars believe them to be descendents of a former Jewish kingdom in southern Arabia (now Yemen) forced to move following conquest by an Ethiopian king in 525 BCE.

A group of Jews from the area around Bombay, India, are known as the Bene Israel, and they trace their origins to Jews who fled persecution in Judea during the second century BCE. They observe some of the ancient dietary laws, but have adopted some Hindu practices as well, including not eating beef and prohibiting widows from remarrying. The Cochin Jews of South India appear to represent an independent colonization event in India. Their tradition describes their migration as dating from Solomon's time in the tenth century BCE, although the earliest evidence of their presence in India, inscriptions on two copper plates, dates to about the year 1000. Many of the Bene Israel and Cochin Jews moved to Israel during the latter part of the twentieth century. There is also a long history of Jews in China, going back at least as far as a Persian business letter written in Hebrew in 718, found in Xinjiang Province. About a thousand Jews settled in Kaifeng, midway between modern Beijing and Shanghai, in the tenth century, where they were known as "scripture teachers." The practices of the Kaifeng Jews were consistent with those of mainstream Judaism, focusing on the Torah, though there were some adopted practices reflecting local Chinese culture, including burning incense in honor of biblical figures as well as Confucius. The Kaifeng Jews are now largely assimilated out of existence; however, there are more recent Chinese Jewish communities resulting from refugees fleeing Russian pogroms around 1905, the Russian revolution of 1917, and Hitler's attempted genocide in the late 1930s.[10]

The Coptic Christians provide another example of preservation of an early set of religious practices maintained into modern times without the changes experienced by various other groups. In about 50 CE Jesus' disciple Mark, the namesake and author of the New Testament Second Gospel, carried Christianity to Egypt. A small community of Christians first developed in Alexandria and spread throughout Egypt over the next fifty years. The Coptic Christians are of particular interest for religious historians because of their maintenance of the cultural traditions of an early group of Christians, making Christianity's roots in Judaism apparent. For example, Copts observe sabbath on Saturday, circumcise male infants, and observe the dietary laws as outlined in the Old Testament. The Copts formally bid adieu to what became mainstream Christianity in 451 CE when they rejected the statements of the Council of Chalcedon, which declared Jesus to be one person existing in two natures (human and divine; diphysitic). For the monophysitic Copts, the two natures could not be separated. There are about 10 million Copts worldwide, located primarily in Egypt, Sudan, and other African countries.

The Samaritans, Jews of Ethiopia, and Coptic Christians are early offshoots whose religious practices have changed relatively little over time. Both isolation and strict adherence to tradition have preserved their customs. Change, when it does happen, is not limited to the divergence of offshoot groups or to the adoption of outside cultural elements. A great deal of change can occur within traditions over time. For example, Judaism, in all its modern variations, is quite different from the practices of the ancient Israelites.

Convergence

Convergence in biological evolution happens when similar features arise for reasons other than common ancestry. The similar torpedo-like shape of distantly related life forms, such as tuna, dolphins, and penguins are convergent and adaptive to rapid pursuit of aquatic prey. The stunning similarity in organismal form and function between marsupial mice and placental mice, between marsupial flying phalangers and placental flying squirrels, and between marsupial wombats and placental groundhogs are also examples of evolutionary convergence, reflecting similar adaptations for similar lifestyles in distantly related groups of mammals.

Convergences in human cultures also occur. Perhaps the most obvious convergence among religious cultures is in the moral systems they espouse. The Golden Rule or the ethic of reciprocity exists in nearly all religions in some form; a sampling of statements from religious texts follows. "Not one of you is a believer until he loves for his brother what he loves for himself" (Islam; Forty Hadith of an-Nawawi 13). "You shall love your neighbor as yourself" (Judaism and Christianity; Leviticus 19.18). "Whatever you wish that men would do to you, do so to them" (Christianity; Matthew 7.12). "A man should wander about treating all creatures as he himself would be treated" (Jainism; Sutrakritanga 1.11.33). "Try your best to treat others as you would wish to be treated yourself, and you will find that this is the shortest way to benevolence" (Confucianism; Mencius VII.A.4). "One should not behave towards others in a way which is disagreeable to oneself. This is the essence of morality" (Hinduism; Mahabharata, Anusasana Parva 113.8). This similarity is expected to be shared among traditions with common origins such as Judaism, Christianity, and Islam. However, it is not expected to be shared for historical reasons among traditions arising independently, at different times and in different regions, as observed in the set as a whole.

The ethic of reciprocity is regarded by many as the most concise and general principle of ethics, and the human capacity to strive for ethical behavior is adaptive, much as the human capacity for language is adaptive. The ethic of reciprocity allows individuals to work together for the good of a larger group, presumably including kin, free of the competition and strife found outside the group. Individuals can better cooperate in hunting, farming, and defense if they agree to treat each other well. Our nature as social animals, and our capability and concern for ethical social behavior, are instinctual and genetically influenced to at least some degree. For this reason a recurrent set of human traits and social themes, including concern for reciprocity, friendship, group loyalty, family, and agreements or expectations regarding mutual aid, are found in virtually all religious cultures and all human populations. The widespread nature of these behavioral ideals indicates biological features common to all humans, including an evolutionary history for the biological mechanisms by which instincts function.

This points to the paradox of moral systems. Ethical, moral behavior is applied, first and foremost, within groups, and only secondarily between groups. Friendship and group loyalty are often forged and tested in the competition and hostile interactions among groups. As noted by Matt Ridley, religions tend to emphasize "the difference between in-group and out-group: Israelite and Philistine; Jew and Gentile; saved and damned; believer and heathen; Arian and Athanasian; Catholic and Orthodox; Protestant and Catholic; Hindu and Muslim; Sunni and Shi'a . . . There is nothing especially surprising in this, given the origins of most religions as beleaguered cults in tribally divided, violent societies."[11] Richard Alexander concludes, "the rules of morality and law alike seem not to be designed explicitly to allow people to live in harmony within societies but to enable societies to be sufficiently united to deter their enemies."[12]

In previous sections I have focused on cultural features shared by some but not all cultures as a means of determining their historical relationships. Widespread convergence in basic ethical ideals, similar in historically unrelated cultures, tells us less about the history of those cultures and more about the biological evolution of the underlying social behaviors and instincts. Humans are highly social creatures living in partitioned, multi-societal cultural environments. Simultaneous pursuit of peace within groups and conflict among groups can be understood in an evolutionary context. Such an understanding exposes the challenges inherent in actually applying the ethic of reciprocity as broadly as it is espoused.

Adoptions

Organismal evolution involves two broad categories of heritability. Usually, traits of organisms are transmitted vertically from parents to offspring. Rarely, however, traits are transmitted horizontally between members of different evolutionary lineages. This occurs, for example, when viruses transfer foreign genetic material into the genome of their host's reproductive cells, or when a plasmid transfers DNA from one bacterial lineage to another. At the level of species rather than individuals, horizontal evolution happens when members of different species interbreed. In these cases heritable traits, specifically DNA sequences, are introduced from an outside source, and may be transmitted vertically thereafter. Events of horizontal evolution can be shown within phylogenetic trees by inclusion of horizontal, or reticulate, branches connecting distinct lineages. This adds realistic elements of a network to what is an otherwise strictly bifurcating diagram.

The history of religions is saturated with borrowings or forced adoptions of cultural elements (syncretism), and these are analogous to events of horizontal evolution among organisms. The frequency of cultural interchange, whether desired, resisted, or ignored, is one reason why the metaphor of evolution for cultures should be used cautiously. Vertical transmission is the norm for animals, but this is not the case for cultures.

Several stories from Hebrew scriptures, including that of Noah and the Ark, have precedents in earlier traditions of Sumerians, Babylonians, or Canaanites. The ancient Babylonian epic poem *Gilgamesh* is known from a set of twelve clay tablets dating to about 650 BCE discovered in modern-day Iraq. Passages from the story have also been found on smaller tablets dating to about 2000 BCE, and linguists consider the story older still. Both *Gilgamesh* and the Noah legend involve a comprehensive flood of divine origins designed to punish people for their sins. In *Gilgamesh,* a man named Utnapishtim is told by an assembly of gods to build an ark. The gods consider him righteous and want to save him. Utnapishtim is also told to "load the seed of every living thing into your ark." In the Bible story, God tells Noah he is deserving of salvation and instructs him to take pairs of each kind of animal onto the ark to "keep their issue alive all over the earth." Utnapishtim took his family and some friends along with him. Noah took his wife, his sons, and their wives. Both were given seven days to complete their ships, before the rains inducing the flood were to begin. Both released birds to see if they could find land, or if they needed to return to the ship.[13]

The ancient Israelites incorporated a variety of religious concepts, rituals, and even psalms from rival Canaanite sources. Going too far with this was evidently a major concern for early Israelite leaders, given the repeated warnings in Hebrew scriptures against practice of various Canaanite customs, particularly idol worship, as in the legend about Moses smashing the tablets inscribed with the Ten Commandments on seeing his followers worshiping a golden calf.

Intentional omission of traditional practices has also been

common. The early spread of Christianity was facilitated by dropping certain customs from its Judaic origins while adopting others. A new religion is less appealing if converts are forced to accept unfamiliar customs such as circumcision or dietary restrictions, and these traditions were dropped as the followers of Jesus sought greater appeal for their new religion. Similarly, it is difficult to create a new, suddenly popular holiday for celebrating a founder's birth. It would be easier to add the birthday celebration to an existing holiday. This was the case with the decision to celebrate Jesus' birthday at about the time of the winter solstice holiday, when the short days of winter begin to lengthen again. Roman Emperor Aurelian had marked December 25 as *Dies Natalis Solis Invicti* or "the birthday of the unconquered sun" in recognition of the long-popular pagan feast day for *Sol Invictus,* the sun god. December 25 concluded five days of celebration. In 273, shortly after Aurelian made his decision, Christian leaders chose December 25 as the day to celebrate Jesus' birthday. By the mid-sixth century, Roman Emperor Justinian recognized the former solar feast day as an official Christian holiday.

Seventh-century conversion of most residents of Arabia from paganism to nascent monotheistic Islam by Muhammad was eased by the widespread monotheism among Jews and Christians at the time. The concept of monotheism had been established already and was easier to adopt. Mecca had been a center for pagan worship for generations of desert nomads, and Muhammad's success was facilitated by being able to reorient Mecca as a center of Islamic rather than pagan worship following his military conquest of the city.

A few more examples of cultural borrowings help to show its frequency. Several small communities of Jews in southern India follow Hindu tradition in not eating beef and in maintaining a caste system, in which some individuals have no communal rights. Some Ethiopian Jews use prayer mats and bow in the direction of Jerusalem during prayers. "Tangri" is the Turkic language name for a sky god first known from inscriptions on a seventh-century stone pillar from the Orkhon River in modern Mongolia. It survives as a synonym for Allah in modern Muslim Turkey. Some Christians celebrate the Jewish festival of Passover, celebrating the end of Jewish slavery in ancient Egypt.

Manicheism, a religion founded by the self-proclaimed prophet Mani (216–276) in about 240 CE, is a prime example of religious cultural mixing. Mani was born and grew up in Mesopotamia, which already had a long history as a buffer zone between competing Mediterranean and Asian empires, where many different religious cultures coexisted. Israelites, Iranian Zoroastrians, and Mesopotamian astrologists had lived in close proximity there for a thousand years. During Mani's time, the region was populated by their descendents, including followers of numerous Christian sects as well as hybrid Jewish-Christian Gnostic groups. Mani was an ambitious prophet who traveled as far as India, where he learned about Buddhism, and he inspired a network of missionaries at home and abroad. Mani sought to speak the language of local peoples, both literally and figuratively, borrowing ideas, symbols, and religious terms from nearly every tradition he encountered. This makes a new religious message appear consistent with, and possibly a perfection of, older truths.[14] In his intentionally inclusive religion, Mani presented himself as the latest in a line of prophets following Zoroaster, Buddha, and Jesus, and he included significant Iranian cultural symbols, such as Ahura Mazda (the chief deity of the Zoroastrians), in his worldview as well. His success in appealing to a broad diversity of people can be seen in the ruthless persecution of his followers by established religious powers stretching from Rome to China. Mani died in prison at age sixty, although the last active Manichaean community was found in seventeenth-century southeastern coastal China.

Lost Tribes

Many people hold the origins of their religious culture as a key component of their identity. The origins are taken to explain the intangibles of their personal history, including much of what they believe. If the culture's origins are old and broadly accepted, this provides an aura of authenticity and comfort that allows individuals a secure place in their local community, in the broader community of cultures, and in history. This at least helps to explain the allure and frequency of religious cultural borrowing, given the difficulty in establishing entirely new cultures. Cultural identities also influence, and sometimes dictate, the relationships that outsiders have with insiders. Thus, forcing cultural identities on poorly understood or out-of-favor groups of outsiders provides a rationale for transfer of biases, when politically expedient.

This brings us to one of the most famous loose ends in the history of religion. The biblical story of the lost tribes of Israel is one of the oldest missing persons mysteries in recorded history, and has proven to be an irresistible adoption claim and historical insertion point. As noted earlier, the biblical narrative relates that Abraham had eight sons. One of his sons, Isaac, was the father of Jacob, who had twelve sons (Reuben, Simeon, Levi, Judah, Dan, Naphtali, Gad, Asher, Issachar, Zebulun, Benjamin, and Joseph), who in turn gave rise to the twelve tribes of Israel. The twelve tribes were divided by civil war into two kingdoms. The northern kingdom of Israel included ten tribes, and the southern kingdom included the tribes of Simeon, Judah, and part of Benjamin. The northern kingdom was eventually brought down by the invasions of the

Assyrian kings Tiglath-Pileser III in 732 BCE and Sargon II in 721 BCE, and the ten tribes were exiled to Assyria, Media, and neighboring areas, spanning parts of modern Iraq, Iran, and Azerbaijan. A portion of the southern kingdom was exiled to Babylon in 586 BCE and another portion remained. The ten exiled northern tribes are presumed by most historians to have been assimilated into the Assyrian population or killed, as had many others who had been conquered by them. There is archaeological evidence showing that individuals with Hebrew names were enlisted in Assyrian army units in the seventh century BCE. That seems to be the last historical evidence for the existence of members of the lost tribes. Hence, this is where the history of the myths surrounding their later appearances begins.[15]

The mythological and imagined reunion of the ten lost tribes of the northern kingdom with the tribes from the southern kingdom was outlined in scriptures (e.g., Ezekiel 37:16, Isaiah 11:11–12) and became linked to the final redemption of Israel. This tale was carried far and wide by proselytizing Christians, so its popularity as a means for understanding unknown peoples and influencing perceived origins is not surprising. The mantle of common descent is readily adopted or imposed, and, until the recent advent of genetics, difficult to reject.

Surprising to those who doubt there is any historical relevance in biblical narrative, there is some recent molecular genetic evidence supporting the historical division of the tribes based on the hereditary nature of the priesthood, the Cohanim. An analysis of Y-chromosomes, present in males only, from contemporary Jews in Israel, Canada, and Britain, show a close evolutionary relationship and strikingly greater similarity among all self-identified Cohanim, regardless of whether the individuals sampled are Ashkenazic (from northern Europe) or Sephardic (from northern Africa or the Middle East). The estimation of dates from the molecular data are also generally congruent with those based on archaeology and biblical accounts. The distinctive, molecular features for Cohanim Y-chromosomes are dated to an origin around 650 BCE, although the potential for error in the estimates remains large.[16] The close relationship among Cohanim stems from adherence to the tradition for marrying only within the group.

Many have claimed descent from the lost tribes of Israel, from antiquity to the present. Conjecture has linked the lost tribes to a kaleidoscopic array of groups from nearly every continent, with Arabia, India, Ethiopia, and the Americas figuring prominently, and including the Nestorians of Mesopotamia, the Afghans (many of whom call themselves "Beni-Israel"), the holy Shindai class of Japan, and the high caste of Hindus.

Let's consider a sampling of cultural adoption and imposition surrounding the lost tribes mythology. The popular medieval fiction *Mandeville's Travels* was written about 1366 under the pseudonym of Sir John Mandeville. The still-unknown author concocted a travelogue with a free-flowing mix of Jewish and Christian legends and claimed to have found the lost tribes in the mountain valleys of a distant land beyond Cathay (China). This fanciful adventure story fed a hunger for reports of distant lands and peoples and helped perpetuate the lost tribes mythology. The Venetian merchant Marco Polo (1254–1324) also wrote of Jewish kingdoms in the distant East. Mandeville's book achieved much wider circulation than did Marco Polo's at the time, based on its translation into more languages and an abundance of surviving medieval manuscripts.

The Anglo-Israelism movement, beginning in the eighteenth century, posits that modern Anglo-Saxons are the descendents of the lost tribes of Israel, via the Scythian tribes (from present Ukraine), and that Britain and its affiliates, including the United States, have inherited the covenant of Abraham, making them God's chosen people. The movement gained a father figure with publication in 1840 of *Our Israelitish Origin* by the Christian phrenologist John Wilson. Wilson's arguments did not gain broad support, though they did resonate with some, including C. Piazzi Smyth, the Astronomer Royal for Scotland, who managed to deduce from the measurements of the pyramids in Egypt that the English were descended from the lost tribes.[17] Anglo-Israelism's proponents developed the theory that, bolstered by the lost tribes' covenant with God, the Anglo-Saxon race would achieve world supremacy. As Japan rose to power in the nineteenth and twentieth centuries, arguments arose that Japan, not America, was the true sister tribe to Britain in an Anglo-Japanese-Israelite constellation.

A Japanese-Israelite connection and claim was published in 1875 by Norman McLeod, a former herring fisherman and missionary from Scotland. One of the first Japanese to propose common ancestry with the lost tribes was Saeki Yoshiro (1871–1965). In an appendix to an academic book on Nestorian Christianity published in 1908, he claimed Jewish ancestry for the Hada clan, an Asian mainland group arriving in Japan during the fifth century and settling in Uzumasa, in the vicinity of Kyoto, though the evidence offered on shared word roots in support of the claim was weak.

In North America, the mythology of the lost tribes was expanded by the Mormons. Founder Joseph Smith (1805–1844) was born to a Methodist Christian family and spent most of his childhood near Palmyra, New York, in an environment of fervent Protestant revivalism. In 1820 he had a divine revelation, and two years later claimed that he had discovered and translated, with divine assistance, gold tablets inscribed with the words of Abraham's God. He published the translation in 1830 as the Book of Mormon. The Book of Mormon recounts the early history of peoples in America from about 600 BCE to about 420 CE. There were two distinct groups of colonists. The Jaredites came first, following a 344-day-long oceanic voyage, and were eventually displaced by the second group, led by an Israelite named Lehi and his followers, who were primarily descendents of Joseph, the leader and namesake of one of the lost tribes of Israel. Their remaining descendents are said to be represented by the North American Indians. Mormon beliefs about dispersal of peoples extend to the Pacific as well. According to one Mormon source, some Israelites departed the west coast of North America about 55 BCE, colonizing and leaving descendents in New Zealand, Samoa, Tahiti, Tonga, and other Pacific islands. Mormon writings insist that the followers of Joseph Smith will eventually be joined by the remnants of the lost tribes "in the literal gathering of Israel and in the restoration of the Ten Tribes," and "that Zion (the new Jerusalem) will be built upon the American continent."[18]

Also in North America, the Church of God and Saints of Christ was founded in 1896 based on the claim that the lost tribes were the ancestors of black people. A rival group, the Commandment Keepers, claimed the black peoples were descended from the tribe of Judah, whereas the lost tribes were the white race.

The recent tragic history of conflict among the Tutsis and Hutus in the central African nations of Burundi and Rwanda have also revived an adopted link to the lost tribes, in this instance for the Tutsis. Tutsis had long been linked to the lost tribes of Israel or other nonindigenous peoples by missionaries or colonial authorities because of physical differences—being taller and having more angular features—and because of their tendency to hold positions of higher social status relative to the more populous Hutus. Occasionally, their imagined link to the lost tribes was anti-Semitic in origin and used to characterize them as greedy and rootless. Following massacres in 1994 in which hundreds of thousands of Tutsis and moderate Hutus were killed, the notion that the Tutsis might be descendents of the lost tribes brought hope to some of greater international

Figure 5.9 (a) Phylogeny for human populations and (b) map summarizing the historical migrations of humans across the continents with approximate dates.

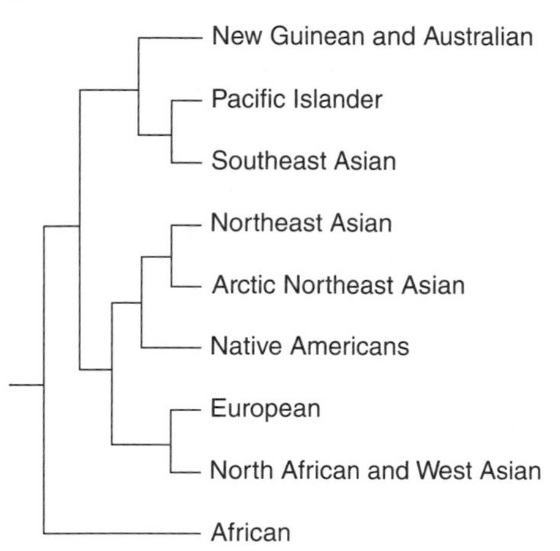

b.

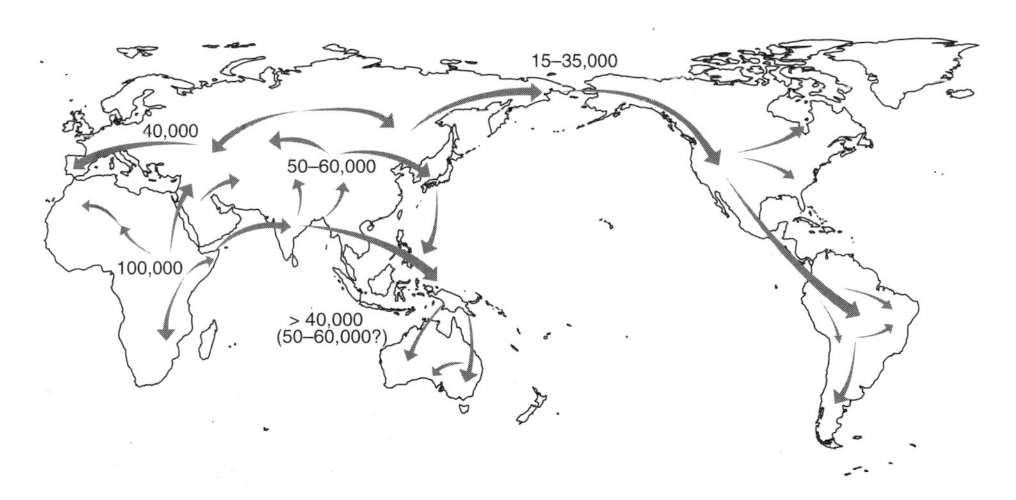

assistance and eventual redemption. Interest in a Tutsi-Israelite identity reflects similar movements among the Shinlung in the eastern states of India, about 300 of whom have emigrated to Israel, among the Baluba in Zaire, and the Lemba in South Africa.

The circumstances of the Lemba are particularly interesting, in light of their syncretistic religious practices. Some belong to Christian churches, some are Muslims, and others claim to be Lemba by religion as well as ethnicity. Nineteenth-century missionaries wrote about their "jewishness" in religious practices as well as their lighter skin. The Lemba are monotheistic, observe strict Old Testament dietary laws, and have ritual male circumcision. Further circumstantial evidence includes reports from a German missionary in 1867 of Lemba beliefs that "God had made the man from the same material as the stones and then his wife. Then he told them to multiply themselves. All people were killed once by water, the sun was dark and there was a great flood."[19]

Can evolutionary biology be applied to these longstanding and deeply felt debates on the origins of peoples, including the fate of the lost tribes? Most such debates are safe from the dispassionate eye of science; however, a few can be addressed, at least in part. Study of the evolutionary genetics of the Lemba tribe in parts of Zimbabwe and South Africa do provide some support for their claims. Molecular markers (microsatellite haplotypes) for 399 Y-chromosomes (males only) have been compared among Lemba, Bantu (central and southern Africa), Yemeni, Sena, Jewish-Sephardic, and Jewish-Ashkenazic population samples, and the Lemba Y-chromosome traits are clearly divided into Semitic (Yemeni, Sena, Sephardic, Ashkenazic) and Bantu clades. This includes significant sharing of some genetic traits, at high population frequencies, among the Lemba, Jewish-Sephardic, and Jewish-Ashkenazic samples, which are present at only low frequencies in Yemeni and Arab population samples.[20] This is at least consistent with the oral tradition of the Lemba, in which they claim descent from Jews who came to Africa from "Sena," which is variously identified by them as being in Yemen, Judea, Egypt, or Ethiopia. The close relationship between Semitic and Bantu clades indicates interbreeding among them, as might be expected.

Extensive molecular analyses of the relationships among primary human geographic populations have been completed, and there is no evidence supporting the claim in the Book of Mormon that there is a connection between North American Indians and Jewish ancestry. Instead, the evidence, from genetics as well as linguistics and archaeology, all support the close relationship between North American Indians and Asians. Phylogenetic analyses of maternally inherited mitochondrial DNAs indicate that North American Indians are most closely related to Asians, specifically populations from Siberia, with their migration to the New World having taken place over 14,000 years ago.[21] Studies of paternally inherited Y-chromosomes substantiate these same conclusions.[22]

Uses of Evolution and Limits of Metaphor

Is evolutionary biology useful in understanding religious cultural variation and change? Yes, I think so. Evolutionary understanding helps to free religions of the burden of literal interpretation of ancient biblical legends that require physical impossibilities. Examples include creating a woman from the rib of a man, a virgin giving birth, and the collecting and loading of representatives of all life forms on a smallish wooden boat. By definition, literalists cannot pick and choose when to be literal. In recognition of this benefit, many religious groups have long since dropped insistence on literal interpretations of their ancient legends.

Understanding evolutionary biology frees people to see

religious cultures as originating and changing as a result of human, rather than divine, history. Doing so puts greater responsibility on people for the outcomes of religious cultural activity, including the excesses. "Holy wars" and claims of superior status for particular groups can be deprived of divine sanction. Potentially, so-called eternal disputes could simply become old history.

Evolutionary analyses of human populations show that all humans are in fact a single large family whose ancestors lived in and dispersed from Africa about 200,000 years ago. Analyses also show human genetic diversity to be as great or greater within populations than it is between them.[23] These findings are consonant with some of the most valuable teachings within religious cultures, the teaching of shared human origins and sensibilities. As noted above, the ethic of reciprocity, the Golden Rule, that is espoused by all cultures may be practiced more often when the common history and culture of humans is better appreciated.

Phylogenetic methods developed by evolutionary biologists for reconstructing historical change among organisms may be applied to estimate historical relationships among languages where suitable data on cognate words exists. Though the mechanisms of change differ, some analytical methods can be shared. Diagrams summarizing historical relationships among entire cultural traditions can also be inferred. Depicting historical relationships for cultures and their features makes it possible, in some cases, to identify which traits are shared because of common cultural origins, which are shared because of cultural borrowing, and which have arisen independently. These diagrams making shared histories explicit call attention to the common threads connecting diverse cultures, whose differences often manage to get more attention.

Understanding biological evolution also helps define the limits for use of evolution as a metaphor. Although a near-infinite set of parameters and interactions are involved, organismal evolution stems from the differential reproductive success of individuals varying in their heritable traits. That evolution is used metaphorically in describing small-scale variation and change in cultures, as outlined in this chapter, follows from observing that features such as alternatives in word choice and ritual performance are not directly heritable. Thus the importance of distinguishing metaphorical from literal uses of evolution, in the biological sense, pertains to understanding mechanisms of change. The limits of evolutionary metaphor are reached prior to identifying mechanisms of change. The reason for making the limit explicit is to promote understanding of cultural change, including religious cultural change, as stemming from human politics and history, and to no small degree, to free evolutionary biology from misguided perceptions that minor variations on cultural themes are genetically determined and directly sorted by natural selection due to some adaptive advantage conferred on practitioners. E. O. Wilson has remarked, "The genes hold culture on a leash. The leash is very long, but inevitably values will be constrained in accordance with their effects on the human gene pool." This implies broad genetic constraints on culture, and this makes sense in speaking of the capacity for culture and its basic outlines as shared by diverse groups. However, it does not apply in speaking of the variant aspects of cultures, falling within the area circumscribed by the very long leash. The real promise for applying biological evolution to study the origins of common features for cultures will come from studies of the evolution of brain form and function, and the particular constraints they place on cognition and social behavior, a task well beyond the scope of this book.

If it come to prohibiting, there is not aught more likely to be prohibited than truth itself; whose first appearance to our eyes, bleared and dimmed with prejudice and custom, is more unsightly and unplausible than many errors.

—John Milton, *Areopagitica* (1644)

6

THE ROLE OF EVOLUTION IN COURT AND CLASSROOM

Our minds have evolved in natural environments where problem solving and fact discovery have been essential to survival. Throughout human history, we have continually probed our environment to understand its physical processes. Our current understanding of evolutionary processes allows us to explain and even predict aspects of the natural world by using observations and reasoning. By discovery of the facts of nature, we have made countless chaotic situations more comprehensible. The urge to understand how nature works is both practical, being relevant to health and survival, and deeply satisfying.

When scientific discoveries about nature conflict with cherished supernatural beliefs, however, conflicts may arise. Most of us are relatively open to an unwelcome discovery if there is irrefutable evidence in favor of it. If you find documents stating that you are adopted, you may at first deny it, but then you gather genetic evidence. If the evidence shows unequivocally, in multiple independent tests, that you are adopted, then you adjust your understanding, however difficult that may be. Those with literalist beliefs in the ancient creation legends of their religion may find it difficult to accept the scientific findings of evolution. Some seek protection from the unequivocal finding of common descent for all life forms, including humans, through the legal system. For example, attempts have been made in the United States to restrict the science curriculum in high schools by court order. Resisting unpopular discoveries by restricting teaching is, of course, nothing new (see Chapter 1). Education policy in the United States is subservient to public opinion, and has, for the most part, been determined locally rather than nationally. The history of legal rulings on education is interesting in terms of social and cultural change; however, it is not bound to keep pace with scientific discovery.

Given the overlapping concerns of science and religion in understanding life on Earth and the alternative approaches to understanding that are used, conflicting views are inevitable. Science succeeds via continual hypothesis testing and the accumulation of evidence, whereas religions thrive on faith in traditional beliefs. The First Amendment to the U.S. Constitution, adopted in 1789, has become a key piece of legislation in the United States for resolving disputes over teaching of evolutionary biology in high schools. The First Amendment guarantees that "Congress shall make no laws respecting an establishment of religion, or prohibiting the free exercise thereof." The statement is sufficiently broad to allow multiple interpretations. Those who want to teach religious views in science classes invoke the "free exercise" clause, and those defending unfettered science curricula and teaching of evolution invoke the "establishment" clause, against establishment of religion. This amendment was intended initially to prevent any particular religion from being mandated by government (as seen by the amendment writers in the Church of England) and to prevent the persecution of the followers of other religions. It is not surprising that science has flourished under these conditions, where democracy and religious freedoms have been prized.

Legally ensuring the respective freedoms for science and religion has not been easy. Despite frequent challenges, however, the long-term trend in the U.S. courts regarding science teaching in public schools has been to keep high school *science* curricula free of religious accounts of the natural world. Religious accounts can be and are reviewed in various other classes, including those on sociology, comparative religion, and world history. During the 1920s, teaching of evolution was banned in schools in some states where religious fundamentalists felt the biblical account in Genesis to be under attack. In the famous *Scopes* trial of 1925, John Scopes, a high school biology teacher, was found guilty of teaching evolution, including man's common ancestry with other species, as outlined in his textbook, by the Tennessee Supreme Court. The fact that his penalty for conviction was a mere $100 showed the high degree of ambivalence for this law, which nonetheless remained in effect until 1967. During the 1960s and 1970s, bans against teaching evolution in several states were rescinded. In the 1970s, fundamentalist Christians sought again to assert their power and require teaching of so-called creation science, though this was ultimately ruled to be a religious account of nature and not scientific.

Recent attempts to mandate teaching of intelligent design are failing to get the desired legal backing as well.[1] Proponents of intelligent design claim that some features of life are too complicated to have evolved naturally. Some emphasize that the inferred designer is not necessarily a deity, others do not. Some intelligent-design proponents accept scientific estimates of life's origins over 3.5 billion years ago and common ancestry for nonhuman organisms, whereas others do not. Regardless of those differences, intelligent-design proponents do agree on the existence of a supernatural designing entity. And this is precisely why intelligent design is not scientific. Supernatural entities cannot be tested scientifically (if they could, they would not be supernatural), and acceptance of the supernatural rests on faith rather than tangible evidence. Thus, intelligent-design activism falls into the category of cultural campaign rather than scientific hypothesis-testing.

The role of evolution and its cultural acceptance can be viewed in two ways. Most often, cultural acceptance of evolution is assessed by examination of court rulings. We tend to look at the small number of cases in which attempts are made to compromise the teaching of evolution and to force religious accounts of biological nature into the science curriculum. A more telling assessment of cultural acceptance is to look at the uses of evolution, including its basic concepts and methods, within the legal and the public education systems themselves. Are there applications of evolutionary biology that aid the pursuit of justice and enforcement of ethics? Yes, a good case can be made. Are there applications of evolutionary biology that aid the pursuit of academic excellence and scientific discovery? Yes, again. Many of the arguments made in earlier chapters provide support for this idea. Legal protection for science education free from the imposition of supernatural explanation is vitally important for understanding the world and promoting our own health and survival. The evolutionary genie is out of the bottle and cannot be put back in.

The U.S. legal system and the curriculum in public high schools are, for better or worse, showcases for our cultural values. The legal system is where we attempt to define and enforce ethical behavior in the broad sense, and the public high school curriculum is where we attempt to provide the cultural background and tools needed for an educated citizenry. In this chapter I will examine the role of evolution in the cultural arenas of the legal system and public schools and universities, asking about its use and integration.

Evolution in Court

Admission of Evidence

How is science used in the U.S. legal system? What qualifies as scientific knowledge? When and how are particular scientific methods deemed reliable and relevant to a given case? These are longstanding, important questions whose answers have changed as the legal system comes to grips with evolving scientific concepts and technologies. The 1923 District of Columbia Court of Appeals case *Frye v. United States* set "general acceptance" within the scientific community as a prerequisite for admitting scientific evidence in court. Though widely adopted and still widely invoked, many have felt this ruling to be too restrictive, because it often results in the barring of testimony and evidence using scientifically compelling but novel concepts. In 1975, a more liberal standard for admissibility of scientific evidence, the Federal Rules of Evidence, was passed into law by the U.S. Congress for federal courts. Rule 702 concerning admissibility of expert testimony stated, "If scientific, technical, or other specialized knowledge will assist the trier of fact to understand the evidence or to determine a fact in issue, a witness qualified as an expert by knowledge, skill, experience, training, or education, may testify thereto in the form of an opinion or otherwise." In 1993, the U.S. Supreme Court sought to resolve the conflict between the more and less restrictive rulings in the landmark decision *Daubert v. Merrell Dow Pharmaceuticals.*

The majority opinion in *Daubert,* written by Justice Harry

Blackmun, stated that Rule 702 of the Federal Rules of Evidence did indeed supplant the more restrictive *Frye* ruling. Specifically, "'[G]eneral acceptance' is not a necessary precondition to the admissibility of scientific evidence under the Federal Rules of Evidence, but the Rules of Evidence—especially Rule 702—do assign to the trial judge the task of ensuring that an expert's testimony both rests on a reliable foundation and is relevant to the task at hand." This decision assigns primary responsibility to the judges to ensure the reliability of scientific evidence. Judges are directed to determine reliability of evidence based on validity of the scientific methods involved, not on general acceptance of the conclusions generated. The opinion reflects views expressed in several "friend of the court" briefs filed by scientists, physicians, and historians of science. Justice Blackmun drew heavily on these, citing Karl Popper, a noted philosopher of science, in saying the science presented in court must be testable. The *Daubert* opinion repudiates adoption of any "definitive checklist or test" for deciding on the reliability of testimony by scientific experts. However, the opinion does provide guidelines. It asks whether the theories and techniques employed by the scientific expert have been tested; whether they have been subjected to peer review and publication (with a warning not to use this as a barrier); whether the techniques employed by the expert have a known error rate; whether they are subject to standards governing their application; and whether the theories and techniques employed by the expert enjoy widespread acceptance. The Court noted that although peer review does not correlate with reliability, the process of peer review "increases the likelihood that substantive flaws in methodology will be detected."

Much has been written about the similarities and differences between science and the practice of law. For example, both science and law are seeking truth, but scientific findings are tested continually in looking for general understanding of nature, whereas the legal system is looking for a quick determination of truth in a particular case and prefers not to revisit the issues. It is no secret, but rarely admitted, that one of the two advocacy groups in any given legal case can be wholly disinterested in the truth—to the point of seeking irrelevant technical reasons for suppressing evidence. This is counter to the best practices of science. Because legal evidence can be the enemy and can make the difference in winning or losing cases and money, the issues surrounding *Daubert* will remain contentious. The current onus on judges as gatekeepers for admissibility of science has and will continue to result in a variety of contradictory decisions, depending on the acumen and background views of different judges. There is always recourse to appeal. Despite the inevitable disagreements in its implementation, the *Daubert* decision and its predecessors have clearly signaled the courts' intention to incorporate scientific evidence and methods.[2]

Forensics

DNA forensics began with a short publication by Alec Jeffreys and colleagues in 1985 showing that variable snippets of human DNA could provide the molecular equivalent of individual fingerprints.[3] Shortly after, another paper demonstrated that these distinctive molecular profiles could link crime-scene evidence to specific persons, and that they could be used to track family histories by comparing molecular profiles among purported relatives. Suddenly, disputes over paternity, maternity, and extended family membership could be put on firm empirical grounds. Ownership of disembodied parts and spilled bodily fluids could be nailed down with precision. Identifiable DNAs, traced to their owner, have been isolated from a single human hair root found on the floor, lip cells left on a beer can, saliva from toothbrushes, postage stamps, envelope flaps, cigarette butts, and chewing gum, as well as semen, blood, urine, and feces.

The methods and concepts are thoroughly evolutionary. Evolution, as descent with modification, predicts that some rapidly changing characters, such as hypervariable DNA sequences, will be unique to all individuals, with the single exception of identical twins, and that more closely related individuals will share more recently derived characters than will distant relatives. Another way to state this, making the evolutionary connection explicit, is that DNA forensics is based on the identification and comparison of homologous genes in different individuals. Homologous similarity is a result of common descent. The alternative form of similarity, convergent similarity, results from chance events or natural selection and is inappropriate as a guide to relationships and identification. The only way to discriminate between homologous and convergent similarity is in the context of evolution, specifically phylogenetic analyses.

As the comparison of DNA sequences among individuals becomes increasingly common, data banks of DNA profiles are growing. These are proving helpful in solving rape crimes. Most states have a statute of limitations, varying from five to fifteen years, for prosecution of rape criminals, instituted largely because of the difficulty in remembering events and faces over long time periods. Now rape cases are being kept open longer. Prosecutors indict unidentified rapists' DNA profiles as determined from semen collected from the victims. This is a way to keep these cases alive, in the hope that an eventual match to the perpetrator will be found and the right person will be held accountable. In 1996, an unidentified Wisconsin rapist's DNA profile was indicted in court and finally matched in 2001 to a prisoner who had been convicted of armed robbery. The accused was then convicted of this past rape. He appealed, arguing the arrest was invalid because the arrest warrant issued at the time of the indictment did not include his name. The Wisconsin appeals court upheld the conviction, holding DNA to be the best means of identification available. Prosecutors in other states have started indicting DNA profiles, as proxy for their unidentified owners, from past rape cases, starting with those that are closest to expiration under the local statute of limitations.

DNA analyses are most clearly interpreted in demonstrating innocence. Innocence is demonstrated conclusively by nonmatching sequences, whereas an interpretation of guilt must inevitably be probabilistic, no matter how small the odds are of some other person having the identical genetic profile. It is sobering to note the efficacy of DNA analyses: To date, about 40 percent of persons arrested for crimes after which DNA evidence was collected have been cleared of wrongdoing. As with other scientific discoveries, useful applications of DNA technology bring ample opportunity for errors and misapplication. Vigilance against errors must be continual, and opportunity for retesting must be ensured. Since 1987, over 140 persons wrongly convicted of rape or murder have been determined to be innocent and freed on the basis of new DNA evidence that was not available at the time of their original trial.[4] If the courts did not recognize DNA as the material basis of evolutionary change, with every person showing some unique differences, these life-changing and justice-serving tests would not be admissible.

Let's consider some specific cases in which DNA evidence and evolutionary principles have been used in seeking to exonerate or convict accused individuals. On a balmy spring night in 1989 a twenty-eight-year-old female jogger was brutally raped in Central Park, New York City. She was found unconscious, with a fractured skull, after having lost about 75 percent of her blood. She was not expected to survive. Police had taken a group of five teenagers, fourteen to sixteen years old, into custody earlier in the evening for a series of attacks in the park. Following separate interrogations, all five confessed to participation in the attack on the jogger. DNA evidence was not yet routine and none was presented at the trial. Forensic evidence that was presented included a hair found on one of the defendants which matched that of the victim in appearance, and another hair on the victim which physically resembled that of one of the defendants. All the defendants were convicted in juvenile court and sentenced, variously, to five- to fifteen-year prison sentences. Early in 2002, a man who was serving a life sentence for other crimes, Matias Reyes, claimed that he had attacked and raped the Central Park jogger on his own. The needed DNA profiles were obtained from the original semen and hairs collected in 1989 and from blood samples from all the suspects. All the DNA evidence corroborated Reyes's claim for sole responsibility, and the wrongfully convicted teens were exonerated. The tragedy of the attacks is unchanged, but a more accurate history of the events and determination of guilt was provided by assessment of the historical record supplied by human DNAs.

After being sequestered for 133 days, Judge Ito arranged for the twelve-member jury in the O.J. Simpson trial to take a ride on the Goodyear Blimp and see a live performance of *Miss Saigon* in order to bolster their morale. Consideration of DNA evidence is ordinarily not so dispiriting. However, the trial of the former football player O.J. Simpson on charges of murdering his wife Nicole Brown and her friend Ron Goldman was far from ordinary. It became a high-profile example for arguments about the quality of DNA forensics work (clearly admissible following *Daubert*) and vehement attacks on the people involved in gathering the evidence. The jury was presented with a mountain of evidence, including more genetic samples than had ever been amassed in a California trial before, and its implications were clear. As reported, the evidence failed to exonerate Simpson and strongly pointed to his responsibility for the murders. Forty-five separate blood stains from two different crime scenes were analyzed by two different labs, and all the results were consistent with Simpson having committed the murders. Blood identified as his was found in multiple places by the site of the killings. Blood from Simpson, Brown, and Goldman was found splattered on Simpson's socks collected from his bedroom in his house after the murder, on a leather glove found on Simpson's property, and from the inside of Simpson's chase-scene Bronco. Despite some quibbling over the odds of blood misidentification being one out of hundreds of millions or of billions, the genetic data itself was clear in its implication. The problem lay in doubts raised about handling of evidence (e.g., were latex gloves worn and changed often enough?), alleged contamination of samples, lack of documented precautions in conducting lab work, loss of credibility by some experts under cross-examination, and claims that police officers planted the incriminating evidence. There was lengthy discussion about the possibility that bacteria could have grown in plastic sealed evidence bags with separately packaged blood samples in them, despite the fact that bacterial contamination, if present, is immediately obvious when analyzing the DNAs—and there were no such sequences. Jury members admitted to being worn out and turned off by two months of contentious testimony about DNA analyses and protocols that they did not understand. The accusations and doubts raised by the defense resulted in acquittal by the jury after less than four hours of deliberation. The DNA evidence had been successfully depreciated by the defense and discounted by the jury. Sixteen months later Simpson was found liable for the two murders in a civil trial and ordered to pay 33.5 million dollars in damages to the families of the deceased. He has since moved to another state, where he is protected from having to make any payments. In this particular case, science, which could have resolved the issue of guilt or innocence, never got a fair chance. Perhaps someday it will.

Phylogeny, Pathogen Sources, and Direction of Transmission

Phylogenetic analyses are qualitatively different from similarity tests and matching of DNA characters as used in the cases described in the previous section. Phylogenetic analyses provide an hypothesis of genealogical relationships showing the history of common descent for a set of individuals. Thus, there is more information available in phylogenies, compared to similarity score comparisons. This allows evolutionary change to be tracked as multiple lineages diversify and spread. In a crime involving infectious agents, pathogens can potentially be collected from multiple alleged victims, sources, and reference populations long after the crime, and the pathogens' evolutionary relationships can still be recovered with careful phylogenetic analyses. This would be much more difficult to accomplish with pairwise similarity scores alone, where convergent similarity arising in living pathogens after the crime would more readily confound analyses of evolutionary relationships.

The value of phylogenetic analyses in forensics is well illustrated by the recent and curious criminal case of the *State of Louisiana v. Richard J. Schmidt.*[5] I worked with several other biologists involved in this case to provide DNA analyses and testimony, so I have some familiarity with the details. The uncontested facts are that a gastroenterologist from Lafayette, Louisiana, broke into the home of his former mistress and office nurse late at night on August 4, 1994, and that he argued with her and gave her an intramuscular injection. He claimed it was a vitamin B shot. She claimed it was HIV. She began feeling ill several months after the injection and a blood test in January 1995 revealed that she had become infected with HIV.

She was a periodic blood donor, and based on tests of those previous blood donations, had a clear record without prior infection. She did not engage in any behaviors placing her at high risk for infection, and her sexual contacts over the previous nine years all tested negative for HIV. The physician was a community leader and Vietnam War veteran. He admitted to having a long-term affair with her, but maintained the infection was not his doing.

She went to the District Attorney's office to file charges on learning she was HIV positive. Moving quickly, the DA's detectives obtained a search warrant and proceeded to the accused physician's office, where they seized his record books for blood samples drawn from patients and a vial of blood sitting in the refrigerator in a back room in his office. The presence of the vial was unusual. Patients' blood samples were sent to the lab soon after being drawn and none were routinely stored there. The physician claimed this sample, drawn from one of his HIV-positive patients, was for his own use and research. Was the physician telling the truth? Might this blood sample link the physician to the victim's infection? Phylogenetic analyses of viral DNAs showing little or no relationship between HIV lineages from the nurse and the alleged source (the blood vial seized from the physician's office) could help demonstrate the physician's innocence, whereas a close, sister relationship among those lineages, in the context of an epidemiological sampling of HIV, would be consistent with the physician's alleged role in transmission.

Are phylogenetic analyses better than other, more routine methods of forensic analyses? In some instances, yes. As noted above, analyses based on similarity alone, such as comparisons of genetic fingerprinting data in which restriction fragment patterns for hypervariable DNA sequences are compared, are subject to greater bias from similarity due to convergence or chance (homoplasy). By contrast, phylogenetic analyses attempt to show explicitly the pattern of common descent among samples analyzed, rather than simple similarity. Phylogenetic analyses have the potential to show homoplasious similarity for what it is—convergence—and not to be misled.

In the *Schmidt* case, portions of the HIV env gp120 gene and the reverse transcriptase (RT) genes were specifically chosen to be sequenced in attempting to maximize the phylogenetic information in the data set. The env gene evolves relatively quickly, being selected upon by hosts' immune systems, and is capable of recovering relationships among recently diverged HIV lineages. The RT gene evolves more slowly, due to greater functional constraints, and can provide insight into relatively older divergences. Using both sequences thus provides a broader range of evolutionary rates than either does alone.

Initial phylogenetic analyses were conducted using maximum parsimony and final analyses were conducted using an explicit model for sequence evolution seeking to account for heterogeneity in rates of change across nucleotide sites and across virus lineages. This was done using the maximum-likelihood optimality criteria in a Bayesian context with the program MrBayes.[6] All analyses of the HIV env gp120 sequences from the nurse, the alleged source, and the epidemiological sampling of HIV patients were congruent in showing the HIV sequences from the victim to form a single monophyletic clade, in showing the alleged source to form a single monophyletic clade, and in showing those two clades to be closest relatives (sister taxa) in the context of the epidemiological sample. This is consistent with the accusation that the physician used the blood sample from one of his patients to infect the nurse, although this rapidly evolving sequence provides no information regarding the direction of infection.

The more slowly evolving RT sequences also indicated their close evolutionary relationship, but with an additional and valuable piece of information. Based on RT sequences, viruses from the victim arose from within the clade of viruses from the alleged source. That is, the alleged source viruses are paraphyletic (incomplete as a natural or monophyletic group) unless the victim's viruses are included and nested within that group. This analysis does provide more direct evidence about the direction of infection, with the clear implication that viruses from the alleged source were used to infect the victim. Viral lineages from the alleged source diverged prior to divergences among the victim's viruses. This difference from the tree topologies based on env sequences can be traced to the slower rate of RT sequence change, suitable for showing an earlier set of divergence events, as expected.

The phylogenetic analyses and rationale above are mainstream among evolutionary biologists; however, phylogenetic thinking and explicit use of evolutionary trees to track genealogy and transmission of virus or bacterial strains between individuals is not yet common in the U.S. courts. The case of *State of Louisiana v. Richard J. Schmidt* set precedent in this area. There was a pre-trial admissibility hearing for the proposed use of phylogenetic analyses in this criminal trial accusing Schmidt of attempted murder. Despite the efforts of the defense to block their admissibility, based on arguments that the viruses were evolving too rapidly to allow tracing of their shared ancestry (they were not), the judge ruled that phylogenetic analyses did meet judicial standards of admissibility as described under the *Daubert* and *Frye* rulings: They were subject to empirical testing, published in peer-reviewed sources, and generally accepted within the scientific community. To further strengthen the analyses and demonstrate their repeatability, DNA sequencing on overlapping sets of viruses was conducted independently in two different labs, at the Baylor College of Medicine in Texas and the University of Michigan. Results from the two labs were entirely consistent with each other.

Though not a panacea, phylogenetic analyses will prove

useful in a range of forensic investigations. As with other molecular forensic approaches they can be particularly effective in demonstrating the innocence of accused individuals. They can also be useful in tracing sources for any transfer of infectious materials, whether viral, bacterial, or protozoan, involving accidental contamination or deliberate infection in personal crimes or acts of terrorism. But these applications are potentially limited by rates of sequence change, which must be sufficiently fast to provide a record of phylogenetic relatedness, but slow enough to preserve sufficient phylogenetic signal before being overwritten with multiple substitutions at individual sites. Further, application of phylogenetic analyses can be complicated by the propensity of some viruses to recombine when representatives of different viral lineages infect the same cell. These possible biases can be detected, however, and addressed as the need arises.

The defendant in *State of Louisiana v. Richard J. Schmidt* was found guilty of attempted murder and sentenced to fifty years in prison, the maximum allowable under the law. The phylogenetic evidence was consistent with the prosecution's case, but there was other evidence that the jury may have found even more compelling. This included the physician having hidden the notebooks of his blood sampling and having a history of threats against the victim as she tried to end their affair. On March 4, 2002, the U.S. Supreme Court rejected an appeal of the verdict without review, thus establishing precedent for use of phylogenetic analyses in U.S. courts of law.

Whether or not all sitting members of the Supreme Court in 2002, including Antonin Scalia, who argued in favor of the creation science position in 1987, were aware of the significance of this case is unclear. However, the role of the U.S. court system has come full circle, and the legal irony is just. The same court system that was used by creationists to suppress the teaching of evolution in the 1920s and earlier has developed a precedent in the case of *State of Louisiana v. Richard J. Schmidt* in which one of evolutionary biology's core principles and methods, phylogenetic analysis, is ruled admissible under *Daubert* and *Frye.* Such an analysis can now be used in the courts as a scientific arbiter of facts about descent with modification, in this case for viruses.

There had been some earlier applications of phylogenetics in criminal cases in other countries. For example, phylogenetic trees for HIVs were used in a Swedish rape trial to see whether or not the viruses circulating in the victim and accused were closely related. They were. In Denmark, HIV phylogenetic analyses linked an infected wrestling coach with a twelve-year-old child who had accused the coach of sexual abuse. The virus analyses supported the child's allegation. An earlier investigation in the United States using phylogenetic analyses indicated that an HIV-positive dentist infected seven of his patients, although the case never made it to trial, being settled out of court.[7]

Knowledge of Evolutionary Variation in Life Histories

Knowledge of variation in species life history traits and ecological succession can also provide crucial evidence in criminal trials. Our understanding of these variable traits of organisms and our ability to use them effectively in court derives from understanding the factors and mechanisms influencing the evolution of populations and species. These approaches are applied particularly well to insects because of their diversity in life history traits, their habitual search for and use of decaying organic matter, and their ubiquity.

An organism's life history is its pattern or "strategy" for growth, survival, and reproduction. This includes many details of the species biology, including rates of growth at various stages of life, changes in appearance over time, capability for dispersal, age at which reproduction begins, frequency of reproduction, numbers of offspring produced per reproductive event, average life span, mating behavior, and parental care. All these traits have evolved under the influence of natural selection, which favors traits maximizing reproductive success of individuals. The broad diversity of life history traits in insects stems from the fact that their environments are extremely variable across locations and over time. Further, the relatively small size of insects facilitates a high degree of habitat partitioning among species, whose specialized uses of resources correspond to high diversity in life history traits both within and between species.

Ecological succession, discussed briefly in Chapter 4, is the change in species constituency for plant and animal communities over time. This process of succession or turnover among species within communities results from progressive modification of the physical environment and is predictable in its sequence. Successional changes in species composition correspond to species differences in life histories. Species adapted to rapid colonization of new habitats and fast rates of reproduction will generally appear more often in earlier successional stages than species that disperse, reproduce, and grow more slowly.

In considering the evolutionary aspects of forensic entomology, it is useful to recognize trends in insect species succession on decomposing bodies. Species succession on human corpses associated with crimes generally begins with necrophagous species, feeding directly on the corpse. These are primarily flies (Diptera) and beetles (Coleoptera). Blow flies and flesh flies in particular are dependent on decomposing matter for food and are fiendishly fast to arrive, often within a minute. As populations of insect species feeding directly on the corpse increase, they attract their own predator and parasite species. These predators often include the burying beetles (Silphidae), the rove beetles (Staphylinidae), and the hister beetles (Histeridae), all of which prey on the eggs and maggots of the necrophagous flies. The parasites attracted to the necrophagous species are primarily the Hymenoptera, which includes ants, bees, and wasps. Many tiny parasitic wasp species lay their eggs on or in the maggots or pupae. On hatching, the wasp larvae feed on the developing fly. Some of these wasps only parasitize a single species of fly, and this knowledge can be useful as an indicator of which fly species were present on a body even after their larvae have developed and left the scene.[8]

Although the links between forensic entomology and the evolution of variable life history and ecological succession traits are often unstated, an understanding of the evolutionary processes involved is important in their application. In fact, such theorics are applied every day in the practice of criminal forensics.

A forensic entomologist must first identify any insect species linked to the crime scene. Life history features (such as length of time between feeding and egg laying) vary greatly among insect species, and an incorrect species identification could provide false evidence. For common species, familiar morphological traits usually suffice. For uncommon species, phylogenetic analyses may be required. Species identification for insects facilitates phylogenetic placement, which allows initial estimates of life history traits. If the insect has the remains of a blood-meal in its digestive tract, the species of the blood donor can be determined phylogenetically as well. If the donor was a human, its genetic profile may also be relevant in solving the crime.

What other sorts of information can forensic entomologists provide in addition to species identification? They can provide estimates of the time of death using several kinds of life history data for flesh-feeding arthropods. These include species succession on a corpse over time, larval weight, larval length, and an "accumulated degree hour technique" that can be extremely precise if the necessary data is available.[9] This latter technique uses recent crime-scene temperature and rainfall records together with biological knowledge about species-specific growth rates at various temperatures to calculate time of egg-laying, based on insect larvae recovered on or near the corpse. After forty-eight hours, forensic entomology is frequently the only reliable method available for estimating time of death. Forensic entomologists can infer whether a corpse has been moved, based on knowledge about habitat preferences of different insects. Wrapping or refrigerating a corpse can be inferred based on finding an atypical succession of insect species. A good forensic entomologist knows the patterns and ordering of flesh-eating insect species appearances, and knows how this pattern varies by geographic region and season. Aspects of these variable patterns are related to adaptive variation among insect species and populations, and an entomologist with an understanding of their evolutionary biology will be able to make informed predictions about species successional variation, even where its details have not yet been described and published.

Complete absence of insects can also provide clues about the sequence of postmortem events, suggesting the body was frozen, sealed in some way, or deeply buried. Victims that were tied up or drugged often urinate or defecate, which may attract species of flies that would not be seen otherwise. Drugs and toxins involved in crimes also can be recovered from flesh-feeding insect larvae. This can be particularly useful if the corpse has been quickly skeletonized by carrion feeders, leaving no soft tissues for toxicological studies.

The following criminal investigation is an example of how our knowledge of the evolutionary biology of insects can be applied.[10] A partially clad woman's body was found near a highway in suburban Washington, DC, in November. The victim had died of multiple stab wounds to the chest and neck. During the autopsy, several large maggots were found exiting the corpse, as is characteristic of fully developed, post-feeding blowfly larvae. Several additional maggots were removed from the neck wounds and clothing. She had been reported missing by her parents eighteen days before the discovery of her body. She was last seen alive in the apartment of the primary suspect, who had a history of sex-related offenses and violent crimes. Hair from the victim's head and pubes was found on bed sheets in the suspect's apartment. Samples of blood and hair taken from the suspect's car also were identified as belonging to the victim. A lot of circumstantial evidence pointed to the suspect; however, an accurate estimate of the time of death was needed to establish the sequence of events. Medical and police examiners offered conflicting estimates of the time of death, varying from two to eight days before the body was found.

The insect larvae observed in and around the body were identified. Climatological data, including maximum and minimum temperature and rainfall, were obtained from a nearby weather service observatory. The largest fly larvae inhabiting the remains were identified as fully engorged third instar larvae and post-feeding larvae of *Calliphora vicina* (European bluebottle blow fly). One specimen showed morphological signs of pupariation (the onset of the larval-pupal transition). Based on the temperature records for the relevant period, the number of days necessary for *Calliphora vicina* to develop from egg to prepupa was calculated. Because the average daily temperature to which the developing flies were exposed was cool (10°C) and because too few larvae were present within the corpse to elevate microenvironmental temperature, the postmortem interval was estimated to be fifteen days, longer than it would be under more typical conditions.

Using this estimate of the fly's life history together with the other evidence noted, the suspect was arrested and charged with first-degree murder, kidnapping, and felony rape. He was eventually tried and sentenced to prison. Later it was learned that he had murdered the victim eighteen days before the body was discovered, and that he had hidden the body in a nearby woodlot by covering it with tree branches and a mattress. Three days later he had transported the corpse to the location where it was found. The materials used to hide the body may have prevented oviposition by adult *Calliphora vicina* during the first few days following death. The fifteen-day postmortem interval estimate was more accurate than that of the police or medical examiners, and it had been instrumental in estimating the time of death, which was used to place the suspect in the company of the murder victim.

Another case, this one involving three murders, also illustrates the application of an understanding of variation in the life history traits of insects. In this Florida triple homicide, one body was found in the back of a pickup truck, another was found in a mobile home, and the third was found outdoors under a tree. The fly larvae collected from each were in different stages of development, and the defense attorneys claimed this supported the murders as unrelated and not concomitant. However, as noted earlier, fly larvae life cycles are temperature dependent. Entomologists working for the prosecution were able to collect the necessary data on microclimate temperatures for the different body recovery sites to establish that the three murders occurred within the same narrow time frame, and this weakened the suspect's alibi.

These applications of evolutionary biology rely on a thorough understanding of variation in life history traits in insects. This understanding must include knowledge of its influences, including phylogenetic, ecological, geographic, and climatic factors, in as much detail as possible. The extent of life history variation found in a given species cannot be predicted without an accurate assessment of its evolutionary history and the genetic, morphological, and behavior constraints that they impose on life history variability. No forensic entomologist, and no judge who understands the field, would claim that life history traits are immutable over time, across geographic regions, or across species. Nor would they attempt to apply that view. Successful forensic entomologists must apply their understanding of the mutability and evolutionary adaptation influencing these natural traits—knowledge found only in the field of evolutionary biology.

To take this one step further, forensic epidemiology can also be considered as grounded in evolutionary concepts and theory. Forensic epidemiology applies evidence about disease in human populations to legal issues of disease causation and control. Epidemiological evidence was significant in lawsuits regarding the use of swine flu vaccine, Agent Orange (an herbicide used by the U.S. military in Vietnam), and Bendectin. Bendectin is a morning sickness drug that was withdrawn from the market in 1983 for allegedly causing birth defects. However, the same active ingredients are making a comeback in the drug Diclectin, as long-term studies failed to support the allegations against Bendectin. The same evolutionary variation noted across species, geographic regions, and habitats noted for life history traits also exists for disease susceptibility in humans. Again, this tacit understanding underlies a great deal of the theory of study design in forensic epidemiology.

Experiments to assess disease causation choose individual human subjects randomly as a safeguard against systematic differences between comparison groups, known as selection bias, which can distort the apparent effect of a treatment. One such form of selection bias is choosing closely related individuals of similar genotypes, predisposing them to react in a particular way to the treatment. Randomization attempts to ensure that each treatment group being compared is representative of the entire population. Forensic epidemiologists must ensure that sample sizes are large enough to detect any actual effects. A large sample is needed to account for naturally evolved variation among human populations in susceptibility to experimental treatments. Effective sampling strategies, designed to remove biases stemming from the evolutionary history of populations—including age, gender, and differences in environmental exposure—are a primary concern in determining the validity of studies whose results are used as evidence in court.

Eugenics

Eugenics provides a cautionary tale about the misapplication of science, with the assistance of the courts, to advance social or political causes. The scope of eugenics was outlined by Francis Galton, Charles Darwin's polymath cousin, in his 1883 book, *Inquiries into Human Faculty and Its Development.* Galton explained, "My general object has been to take note of the varied hereditary faculties of different men, to learn how far history may have shown the practicability of supplanting inefficient human stock by better strains, and to consider whether it might not be our duty to do so by such efforts as may be reasonable, thus exerting ourselves to further the ends of evolution more rapidly and with less distress than if events were left to their own course." In short, the object was to promote the improvement of human stock by selective breeding of superior people. Galton wrote, "If a twentieth part of the cost and pains were spent in measures for the improvement of the human race that is spent on the improvement of the breed of horses and cattle, what a galaxy of genius might we not create!" Lest his commitment be doubted, he states in the final passage of his autobiography, "I take Eugenics very seriously, feeling that its principles ought to become one of the dominant motives in a civilized nation, much as if they were one of its religious tenets." Galton was not shy. He also liked to tell fervent Christians about his analyses demonstrating the ineffectiveness of prayer, in which he showed, statistically, that people who were the subject of frequent prayer neither outlived nor recovered from sickness any faster than others.

Eugenicists sought to increase the frequency of desirable traits and decrease the frequency of undesirable ones in human populations. Sporadic efforts were made to sterilize or ban marriage for individuals deemed unfit, described variously as those with major birth defects, or the retarded or feebleminded. A common high school biology textbook in the United States during the 1920s, *A Civic Biology,* had a section on eugenics identifying mental deficiency, alcoholism, sexual immorality, and criminality as hereditary. It suggested that these defects be addressed by sterilization and segregation of the sexes. Students were encouraged to select eugenically healthy mates. Social policies reflecting some confidence in this approach were implemented in the United Kingdom, Germany, and the United States in the first few decades of the twentieth century. In 1907, Indiana became the first state in the United States to pass a sterilization law aimed at those in prison or psychiatric wards who were considered (by a specially appointed eugenics board) likely to have socially or mentally defective progeny based on examination of the history of illness in their extended families. Similar laws were adopted in thirty other states by 1930. Between 1910 and 1940 approximately 38,000 eugenical sterilizations were performed in the United States.[11] In 1925, a Virginia Circuit Court took on a case, *Bell v. Buck,* testing the constitutionality of its newly passed sterilization law. The law passed and was appealed to the Supreme Court, where Justice Oliver Wendell Holmes, Civil War veteran, wrote in the majority opinion in 1927, "three generations of imbeciles is enough." He went on to address legal precedent with equal delicacy, stating, "The principle that sustains compulsory vaccination is broad enough to cover cutting the Fallopian tubes."

The eugenics movement in the United States began to wither in the mid-1930s for a variety of reasons. Perhaps the severe economic depression of 1929–1931 showed that even the well-born can suddenly find themselves socially outcast. Early reports of Nazi atrocities, including not just sterilization but euthanasia of gypsies, Jews, and homosexuals, likely began to remove the patina of reason from the movement. And what had always been true was finally becoming known: There was no compelling scientific evidence to support the basic claims of eugenics. Prominent among the critics was Columbia University geneticist Thomas Hunt Morgan. Though he kept his reservations private for at least a decade, he stated clearly in his 1925 text, *Evolution and Genetics,* "The important point is that mental traits in man are those that are most often the product of the environment which obscures to a large extent their inheritance, or at least makes very difficult their study." He emphasized that the definitions used for traits such as feeblemindedness were so vague and the number of alternative explanations for them so numerous that no theory of heritability, especially a Mendelian one, could support the eugeneticists' claims of genetic causation and their simplistic remedy.

Moral issues aside, intensive eugenics applied to humans, as managed breeding efforts to emphasize particular traits, would favor problems akin to those in some domestic animals. Most domesticated animals are susceptible to a higher incidence of genetic diseases as a result of inbreeding, reduction of genetic diversity, and obsessive selection for single traits. The brains of domestic cattle, pigs, and dogs are about one-third smaller than their wild progenitors, and many attributes needed for independent survival in the wild are no longer maintained by selection.

The evolutionary concepts of heritability and descent with

modification were invoked, erroneously, as part of the eugenics social movement beginning in the late nineteenth century. Why did it take so long for the claims of eugenics to be discredited? The complementary fields of genetics and statistics that were ultimately necessary to show the false premise of eugenics were still in their infancy in the early twentieth century. Ironically, as an independent scientist, Francis Galton was a significant contributor to the early development of both the fields that would, in the end, fail to lend eugenics empirical support.[12] Genetics and statistics are key disciplines in the study of organismal evolution, and their contribution to our growing understanding of evolution helped discredit the eugenics movement in the United States and Europe. Of course, the issue of human intervention in the future evolution of our species' gene pool has not disappeared. Sterilization and selective abortion continue in various parts of the world, and the debate goes on about the nature of genetic counseling and the subsequent reproductive decisions that are made. However, evolutionary biology is now in a better position to defend itself and the best interests of healthy human populations.

Evolution in the Classroom

What is the role of evolutionary biology in the science curricula of high schools and universities? From the standpoint of nearly all science teachers, evolutionary biology serves as the unifying concept for the life sciences. Evolutionary history is the common crucible through which all life forms have passed, and by scientific investigation of evolutionary processes, students (and researchers) can better understand why and how any biological system works. This means that the role of evolutionary biology, as a set of concepts and methods, is necessary for comprehensive biological understanding. Thus, evolutionary biology is essential to a nearly infinite set of questions about biological systems, spanning all levels of organization. A few examples:

Why is the genetic code structured the way it is?

Why are proteins folded in particular patterns?

Why are cells structured as they are, and why does that structure vary among organisms?

Why do animals and plants develop in the way that they do, and why are there both similarities and differences among the major kinds?

Why is sexual reproduction common among animals but less so among bacteria and plants?

Why do some bird species migrate while others do not?

Why have communities of species formed, and how have they influenced each others' morphology, behavior, and genomes?

What predictions can be made about species' origins, species' extinctions, and the health of ecosystems, given particular changes in the environment?

Testing alternative hypotheses for all these questions will involve assessment of phylogenetic relationships among species and the study of natural selection on variable traits among individuals and species.

To help demonstrate the breadth of application for evolutionary biology and its analytical methods in education and research, consider the following three examples. The first outlines an evolutionary approach to understanding the origins of eukaryotic (nucleated) cells using phylogeny, the second considers the evolution of sexual reproduction using natural selection theory and experiments, and the third is a brief look at applications of evolutionary computation (or genetic algorithms), which mimics evolutionary mechanisms to solve a wide variety of science and engineering problems. These are progress reports rather than completed answers.

I am intentionally linking science education in both high schools and universities with scientific research, as the two are inextricably linked. As a consequence, many educators recognize one of the most important skills to teach is that of critical thinking, meaning the ability to assess the reliability of evidence and its application to hypotheses. This is a skill that will not become obsolete.

Cells

All organisms are composed of cells. Each cell is a marvel of intricate structures working in concert to convert food to energy, which is used in running its affairs and those of the body. Cells bring goods in, ship them out, assemble proteins, replicate and repair their DNAs, divide, and communicate with other cells. How did this bustling mini-domain arise? Is this an irreducibly complicated entity that can only have sprung forth fully formed and without intermediates? Not at all. Phylogenetic analyses show clearly that two organelles within the cells of various eukaryotes, namely mitochondria and chloroplasts, are the result of ancient coevolutionary events. The events entailed endosymbiotic relationships, with bacteria colonizing and living inside the cells of other organisms. An ancient bacterial lineage, related to modern *Rickettsia,* gave rise to mitochondria, and a different bacterial lineage, this one related to modern cyanobacteria, gave rise to chloroplasts. These relationships are well supported based on phylogenetic analyses of homologous genes present in the DNA genomes of mitochondria, of chloroplasts, and of a broad diversity of bacterial and eukaryotic organisms. The genomes of these organelles are much smaller in size than the genomes of their bacterial relatives. This evolutionary streamlining reflects transfer of many of the organellar genes to the nucleus, where some can still be found. That mitochondria metabolize oxygen to create ATP as energy just as their bacterial relatives do and that chloroplasts perform photosynthesis just as cyanobacteria do is further support for their bacterial phylogenetic origins and indicates the benefits that each brought to its host.

Accumulating evidence suggests that the nuclear genomes of eukaryotic cells are themselves combinations of genes from the primary bacterial lineages of Archaea and Bacteria. Genes functioning in cell metabolism are mostly bacterial in origin, whereas genes involved with information transfer and processing are primarily archaeal in origin. Phylogenetic analyses are clear in showing frequent occurrence of lateral gene transfer, in which genes from one organism (or genetic element such as a plastid) move directly into the genome of another organism. The exact patterns of gene evolution, however, including lateral gene transfer, accompanying early evolution of the eukaryotic nuclear genome are difficult to determine, as they test the limits of current phylogenetic methods, and sampling of taxa for homologous, slowly changing genes remains limited.[13] Resolving phylogenetic origins for the separate nuclear, mitochondrial, and chloroplast genomes found within eukaryotic cells is at least a start in understanding the evolution of these complex entities.

Sex

Woody Allen said, "Sex without love is an empty gesture. But as empty gestures go, it is one of the best." He has at least one good point here. Sexual pleasure doesn't require love. People have so many overlapping "issues" with sex and love that attempting to treat them independently, or to claim a single, unambiguous motive, is indeed comical. Understanding the evolution of sexual reproduction in animals is similarly complicated. As with the evolution of cells, some might claim sexual reproductive systems to be too complicated to have evolved in an incremental fashion. So, let's consider a few aspects of an evolutionary approach to understanding sexual reproduction.

Sexual reproduction in animals is defined by two physiological elements. Gametes (sex cells: sperm or eggs) must be made that include only one set of chromosomes rather than two, as found in all other cells, and the haploid sex cells must fuse to make a diploid zygote. In the process, the zygote inherits a sampling of genes from each of two parents, combined in a randomly mixed assortment. This much is understood by many high school students, and it explains why siblings are not genetically identical. Less broadly appreciated are the costs of this genetic recombination system. First, recombination breaks up some favorable gene combinations honed by natural selection—combinations that had served generations of parents well. Second, parents transmit only half their genes to each child, diluting their potential genetic contributions to the gene pool. This also slows population growth because the population of females, capable of bearing young, is halved in sexual reproduction systems proportional to the frequency of males. These costs are not present in asexual species, in which favorable gene combinations are not disrupted each generation and 100 percent of an individual's genes are carried by that individual's progeny. Although sexual recombination is obligate in most animals, it is optional in various other organisms, including many insects, protists, fungi, plants, and bacteria. Despite the costs, sexual reproduction is much more common than asexual reproduction among animals, and this biological question begs to be answered.

Evolutionary explanations must address the advantages of sexual reproduction for individuals and populations and show them to outweigh the costs. Explanations showing the advantage of sex, based on theory, observation, and experiments, may be placed in two categories—spread of advantageous traits and removal of deleterious ones. Regarding the spread of advantageous traits, novel genetic combinations are created by sexual recombination and new advantageous genotypes, even if rare, can be favored and spread via natural selection. Experiments show natural selection to be particularly effective in fixing novel genotypes when environmental change is frequent.

Parasites are a particularly important part of the environment in this regard. If parasites with short generation times adapt to host defenses during the lifetime of host individuals, novel genotypes conferring resistance to parasites in host progeny will be favored.[14] In a study of New Zealand snails exhibiting both sexual and asexual forms of reproduction, the proportion of sexual individuals was found to be positively correlated with the frequency of their parasites. In further support of this evolutionary phenomenon, the parasites in question were demonstrated to be most adept at infecting locally common genotypes of the snail.[15] Comparative analyses showing immune-system proteins to be evolving significantly faster than most other proteins also supports this view.

Evidence for the advantage of sexual recombination in purging deleterious genes is provided in tests of hypotheses known as "Muller's ratchet" and the "deterministic mutation hypothesis." Muller's ratchet holds that in asexual populations of limited size subject to deleterious mutations, the random loss of individuals that are entirely free from deleterious mutations is inevitable. Mutation-free genotypes can be restored via recombination of genotypes in sexual populations, but not in asexual ones. The related deterministic mutation hypothesis, which appears to be a more powerful explanation, holds that most mutations are only slightly deleterious, and it is their accumulation and interaction that serve to reduce viability.

Again, sex can recombine all these genotypes, creating some with more deleterious mutations, which will be selected against, and some with fewer deleterious mutations, which will be favored by natural selection. Theory and experiments show that these mechanisms require rates of deleterious mutations greater than those observed in many wild populations if they are to outweigh the costs of sex on their own; however, the general effect of sexual recombination in increasing responsiveness to natural selection, including avoidance of parasitism, is real and well supported.[16] Thus, evolutionary concepts can help us understand why sexual reproduction is favorable in the long term, despite the costs and despite the fact that alternatives exist.

Evolutionary Computation

Evolutionary computation is a new and exceptionally productive application of evolutionary concepts. The diversity of biological species and their many adaptations for coping with environmental challenges attest to the potency of evolutionary mechanisms, so why not put them to work? The basic idea in evolutionary computation is to emulate the processes of evolutionary biology in solving real-world problems in engineering, industry, and science. Evolutionary algorithms are written for computers to solve optimization problems, in which the best solution is identified from a large set of candidates. The general structure of evolutionary algorithms includes five steps.

1. A population of candidate solutions is generated, and the suitability, or fitness, of each one is evaluated and compared to others.
2. A subset of relatively fit individuals in the population are selected as parents.

3. The parents are subjected to mutational changes, including recombination of components where appropriate.
4. The altered candidate solutions (the progeny) comprise the new population.
5. If the optimality criteria for termination has been reached with current progeny, the process is stopped; if not it begins again with step 2.

Key features distinguishing this approach from other conventional optimality analyses are maintaining variable populations of candidate solutions, selection of a range of fitness solution scores, and imposing multiple alternative mutational strategies. The initial solutions generated in step 1 may be randomly determined or based on a fast heuristic analysis.

But in what real-life situations are these analyses valuable? Satellites are increasingly important for communications, entertainment, weather forecasting, and basic operations in travel, industry, espionage, and defense. Given the greater economy of low-elevation satellite orbits (several hundred miles above ground) and the curvature of the earth, satellites will necessarily lose line-of-sight access to receivers for some part of their orbit. Even constellations of satellites working together will experience losses of coverage. Evolutionary computation was applied to optimize coverage by sets of satellites. It identified unusually asymmetric orbit configurations, with variable distance gaps between individual satellite paths, and these performed better than any configuration found previously with conventional methods.[17] Thus, evolutionary computation has been used to optimize the course and timing of satellite orbits to minimize the frequency of communications blackout.

The world's airlines and major airports in places like New York, San Francisco, Sydney, and London have big scheduling headaches. They must manage aircraft landings, takeoffs, maintenance, baggage routing, staffing, workers' unions regulations, billing, and security clearances, and not just for their own flights and terminals, but in coordinating flight and personnel schedules across cities and countries as well. By repeated rounds of mutating, recombining, and selecting among alternative scheduling solutions, evolutionary computation methods are readily outperforming older scheduling routines and increasing efficiency by as much as 30 percent.[18]

Different applications of evolutionary computation share the five generic steps noted above; however, they may measure different parameters and use different criteria for calculating optimality scores. They may also use different strategies for selecting among individual candidate solutions, different candidate population sizes, alternative severities for mutation imposed on candidates, or variable numbers of algorithm iterations. All these operations have analogous ones in the organismal evolutionary process, which can inform their use in evolutionary computation. Additional promising demonstrations of evolutionary computing have been made in routing data in telecommunication networks, identifying transmembrane domains of proteins based on hydrophobicity estimates, designing electronic voice-recognition devices, scheduling assembly tasks at manufacturing plants, planning exam schedules at universities, designing humanoid soccer-playing robots, picking stocks, predicting currency exchange rates, and, naturally, playing chess. All these applications, and similar ones to come, benefit from a detailed understanding of the variables and mechanisms that influence the evolution of traits and the organisms and populations that bear them. It is our understanding of evolution that has permitted the development of useful processes like evolutionary computation.

Evolutionary biology is taught increasingly as a required, keystone course for university biology majors during their junior or senior years. This is because of evolutionary biology's broad relevance in explaining form and function at all levels of biological organization, and because evolutionary approaches can more effectively unify understanding of diverse biological systems after students have attained some familiarity with the components. High school biology classes and introductory university biology courses also cover evolution topics, though in less detail. Thus, current university biology majors have often had a one-semester course in evolutionary biology following an introduction to evolutionary concepts in other university or high school biology courses. The stature of universities results, in part, from the success of their graduates, and universities are highly motivated to provide training in current scientific concepts and methods. Acceptance of undergraduates to graduate programs in science and to medical schools is based partly on the reputation for excellence in education of the university they attended. For science educators, including those serving on curriculum and admission committees, evolutionary biology prepares students for careers in the life sciences, such as natural resource management, public and environmental health, medicine, and basic research in biology, as well as for teaching scientifically literate citizens.

Most public high school science programs share the same goals as universities, though of course the implementation differs. Public high school science curricula, however, are at much greater risk of interference from activists seeking to introduce supernatural explanations in science classes. Risk is greater there because of the greater opportunity for loss of curriculum control by education professionals. Religious activists elected to state legislatures or public school boards can influence the curricula of public high school much more readily than they can influence curricula at leading public universities, where high academic standards and freedom from academic restrictions are integral to their success and where there is greater national scrutiny.

Creationist activists focus their efforts on changing the high school biology curriculum concerning organic evolution because this is where their particular religious culture and nonreligious culture conflict. Direct religious instruction is barred from U.S. public schools by the First Amendment; thus, creationists attempt to redefine science to include explanations that require supernaturalism and seek to have these explanations incorporated in science textbooks. In the 1970s and 1980s in the United States, some fundamentalist Christians sought to require "balanced treatment" for teaching both natural (evolutionary) and supernatural (creation science) accounts of nature. Creation science, advancing the belief that humans were created by a supernatural being, was deemed to be religious and not scientific by a federal court in 1982 *(McLean v. Arkansas Board of Education)* and again by the Supreme Court in 1987 *(Edwards v. Aguillard).*

Carrying on in the creationist tradition, proponents of intelligent design claim that some features of life are too complicated to have evolved naturally. Although they emphasize that the designer is not necessarily God, proponents uniformly believe that God is the designer, and some acknowledge that intelligent design cannot be empirically tested.[19] Intelligent design has not been able to exclude natural selection as a viable explanation for design origins simply because biological attributes appear complex. Indeed, the appearance of complex, purposeful design is exactly the outcome expected of the natural selection process.[20]

Some intelligent-design activists have dropped the unsuccessful strategy of claiming to be scientific and have adopted instead a "teach the controversy about evolution" approach, based on their view that they have identified serious problems with the theory of evolution. If this were so, nonevolutionary explanations for biological diversity could then be considered. Most public school boards and teachers are disinclined, however, noting the striking similarity between intelligent design and creation science, which the Supreme Court has already designated as inappropriate for science classes.

At heart, the creationist-evolutionist debate stems from cultural conflict. Creationists would like supernatural causation, legitimate within religious culture, to be admitted as a legitimate hypothesis for empirical research, within the culture of science. To do so would run counter to the definition of science, which requires tangible evidence and testable hypotheses. A person's faith can be tested, but the supernatural cannot. Creationists and intelligent-design advocates misunderstand, or intentionally misrepresent, the exciting debate about how evolution happens as being a debate about whether or not evolution happened at all.

The views of creationists and intelligent-design proponents differ considerably among themselves regarding which aspects of evolutionary biology, if any, they are willing to accept. Henry Morris, founder of the Institute for Creation Research, claims the earth is about 15,000 years old, much younger than the estimate of more than 4 billion years based on geology and the physical decay rates of radioactive elements embedded in ancient rock. Morris and other young-earth creationists accept Genesis as a literal account of life's origins, with each species having been specially created by God, with no pattern of common descent. Phillip Johnson, an early proponent of intelligent design, does not tie his creationist, supernatural views about biological origins to a literal interpretation of the Bible, though his statements suggest this is largely strategic. Michael Behe, an intelligent-design proponent and biochemist, accepts that various species have evolved from common ancestors and that natural selection is responsible for most, but not all, of the complex structural adaptations of organisms.

Intelligent-design proponents say that their motivation and

views are not religious. If that were true, it would seem to place the intelligent-design hypothesis closer to the realm of science, where they would like it to be. Yet their own statements show this claim to be false, such as this one from Phillip Johnson: "We are taking an intuition most people have [the belief in God] and making it a scientific and academic enterprise. We are removing the most important cultural roadblock to accepting the role of God as creator."[21] Moreover, Johnson, at least, is not shy about discussing his political tactics: "Our strategy has been to change the subject a bit so that we can get the issue of intelligent design, which really means the reality of God, before the academic world and into the schools."[22] These statements are remarkable for their honesty about the aims of those who promulgate intelligent design. Its advocates are thoroughly mistaken, however, in supposing that supernatural causation in any form can be passed off as scientific explanation.

Community leaders have long recognized the importance of providing modern, comprehensive training in the sciences within their schools. The ability of high school graduates to compete for employment and for admission to selective universities is influenced by the quality of their science education. Where the science curricula of public high schools have been restricted and religious views presented in biology courses, there has been widespread concern about the quality of science education and student preparedness.

No Evolution, No Science

Mainstream cultures in developed countries generally assume a scientific worldview. People want to know the facts, as well as they can be determined, regarding their physical world, which includes everything from the functioning of human bodies to the quality and availability of resources such as food and water, on which our lives depend. Of course, misinformation predominates at times, though where the press is free and accuracy the goal, more factual information should and usually does prevail. The criterion of accuracy as an ideal is shared by investigative reporters and scientists alike, and the public has come to expect reporting that involves science to be as free as possible from biases of ideology. Errors in reporting and the ideological interpretation of scientific findings in opinion columns are common, but can be sorted out and reconsidered over time.

The extent to which a scientific worldview permeates our lives and the integral nature of evolutionary biology to this pragmatic, and widely shared, worldview may be surprising to many. Consider a few items routinely covered in newspapers and on television that make sense in an evolutionary worldview, but do not make sense in a world without evolution. Frequent reports on the need for new vaccines and new antibiotics acknowledge the evolution of resistance by many strains of bacteria and viruses. When viruses that are new to humans are first discovered, such as SARS and HIV, it is widely acknowledged that such epidemics result from our lack of an evolved immune-system defense, and that this is due, in turn, to our lack of prior exposure to the virus. Reports on the dating of fossils using radiometric methods—from 3 billion-year-old bacteria to 1 million-year-old birds—acknowledge the existence of both large time spans suitable for evolutionary change and radical evolution in the form and function of organisms themselves. Development and advertising of the novel properties of crops or produce, including larger strawberries and seedless grapes, acknowledge the ability for selection to produce heritable evolutionary change.

Meaningful discussion of these and many other current issues at the interface of science and society requires, unequivocally, the language and concepts of evolution. Imagine a visit to the doctor's office in which there was no discussion of your family history regarding diabetes, cancers, or other potentially heritable diseases. Imagine an investigation of biological terrorism involving viruses in which the concept of evolutionary relatedness among strains from around the world was not considered. In all such cases, conversations uninformed by evolutionary concepts would be superficial and woefully inadequate as a basis for intelligent decisions.

The technological fruits of scientific inquiry, including airplanes, computers, cars, cell phones, weather predictions, maps of subterranean fossil fuel deposits, satellite communication and imagery, hearing aids, vaccines, antibiotics, kidney transplants, artificial joints, and others are proof that the underlying concepts of math, physics, chemistry, and biology are true. If they were not, the inventions and technology based on them would fail. If a hypothesis or discovery withstands rigorous testing by others, it will eventually be accepted, because it works or because it is true, regardless of the investigator's background, ethnicity, or religion (though it would be naïve to suppose discrimination is entirely absent). This is a beauty of science at its best. It makes no difference what people wish or fear to be true. The processes of nature need not and do not conform to any particular human ideology or preconception. By the same token, scientific findings cannot be fairly suppressed or selectively adopted for ideological reasons for long by those with an open mind. To deny a rational approach for solving a problem in one area, say in explaining human origins, but to accept that same approach in another, say in explaining the origins of coal, is itself irrational. Evolutionary biology is an increasingly well-developed science, comprising a necessary component of a comprehensive scientific worldview. If it had not already been initiated as a discipline, we would have to initiate it now, as the only rational explanation for the observed diversity of life forms and their varied means for living.

Science, law, and religion all lay claim to truths, though their criteria and areas of authority differ. Science seeks to explain the natural world in terms of tangible evidence and material causes. Science continually reevaluates its methods and findings to see if new evidence and analyses support existing views. The legal system seeks truths in resolving particular disputes quickly, and does not routinely revisit its findings. Here, too, tangible evidence and material causes are compelling. Religion seeks and promotes truths about the meaning of human life and the relationship between deity and people. Religious truths are a matter of faith. Religious believers offer their testimony as evidence of that faith, but faith remains a simple assertion.

Science, including evolution, is now quite useful in resolving legal disputes. Religious views have their uses in law too: their tenets have informed legal rulings about ethical behavior. In this light, the boundaries for explanatory power seem clear. Science tells us about the natural world based on tangible evidence, and religious faith may hold dominion over matters supernatural and moral. This is, very roughly, the view of "nonoverlapping magisteria" (NOMA) outlined by Stephen Jay Gould.[23] Science and religion appear to operate in different areas of human experience, so, with an eye toward a more peaceful coexistence, they might be seen as not in conflict. Gould has identified a distinction that can allow "people of goodwill" to agree on boundaries for preferred methods of understanding.

Despite this offer of a truce, some will stray. Passionate creationists are loath to forsake their faith in the literal truth of biblical genesis or, minimally, in supernatural intervention in the natural world, despite the lack of positive evidence. Others, myself included, find NOMA to be more about diplomacy than logic. Morality has its roots in biological evolution, and although science cannot prove the inexistence of the supernatural, if supernatural causation were a scientific hypothesis like any other, it would be rejected over and over again. Interested readers can find plenty of stimulating discussion.[24]

Though I savor cold logic, diplomacy has real social value that can ultimately improve the climate for science by promoting tolerance. Resistance to the fact of evolution by people with strong religious convictions often rests on the perception of threats to their values and way of life. But evolutionary biology, like physiology, organic chemistry, biophysics, or any other study of the natural world, does not justify particular kinds of human behavior. The study of nature shows what is, not what ought to be. Nature cannot be read as a morality play, unless you believe that peregrine falcons should be hunted to extinction for the crime of killing ruddy turnstones in migration. The study of evolution highlights the distinctive capabilities of humans (and a few others) for compassion and ethical behavior. Many people of faith have accepted evolution and more can be shown that perceived threats to their values are not real. Making the attempt is worthwhile; it is likely to reduce social tensions, improve our education systems, and ensure that scientific inquiry remains unrestricted by theology, whether that theology is of a few or of many.

The title of this chapter is "The Role of Evolution in Court and Classroom." For many people this phrase will bring to mind the legal battles over the teaching of human evolution. But this chapter is about much more than the legal battles over teaching about human origins. I could have substituted "biological history" for "evolution" in the title, and that would have worked as well. But I used "evolution" because it is the most appropriate term. Explaining human origins is only one aspect of evolutionary biology, and I am making the case that evolutionary biology in its full range of topics and methods has become useful in our courts and classrooms. The evolutionary nature of the applications of biology are generally underappreciated by the public, however. Many do not yet see that comparative DNA analyses, identification of pathogens' natural reservoirs, and development of vaccines are part and parcel of evolutionary biology. This is not surprising. Evolutionary biology entails synthesis and background knowledge in multiple fields, and single observations of DNA or organisms do not begin to make sense of the whole of biological history. Similarly, it is easier to identify a few constellations of stars in the sky than it is to integrate the wealth of astronomical evidence into a history of the cosmos. Evolution is arguably the most inclusive of the biological sciences, and its reach is expanding as new research areas open.

At the beginning of this chapter I set out to ask whether evolution had been incorporated into two of our primary cultural institutions, public school classrooms and the U.S. legal system. I maintain that not only has evolution penetrated both of these institutions, but more generally modern culture itself. For many, the acceptance of descent with modification among all life forms, including humans, creates no ideological conflict. For others, the benefits of evolution and evolutionary thinking are accepted as pragmatic, useful approaches to real-world problems within the legal system and science classroom, without realizing or acknowledging the evolutionary premises. Those who would deny the tenets and findings of evolution, however, must also deny or ignore a large and growing body of applied evolutionary science.

CONCLUSIONS

In 1783, American statesman-scientist Benjamin Franklin helped finance the first manned flight in a hot-air balloon. The twelve-foot-diameter silk balloon was filled with hydrogen released by pouring corrosive oil of vitriol over red-hot iron filings. Franklin watched as much of the two-hour, twenty-seven-mile flight as he could from his carriage parked by the Tuileries Gardens in Paris. Although some of his colleagues in the Royal Society of London and elsewhere were put off by "ballomania," a seemingly useless indulgence, Franklin was quite keen on the enterprise. When a fellow spectator at the Tuileries Garden asked of what use the new contraption could possibly be, Franklin replied with typical wit, "What is the use of a new born baby?" Beginning with his early experiments on electricity, Franklin demonstrated his conviction that science pursued at first out of simple curiosity might eventually lead to practical uses, unimagined at the outset.[1] Franklin was prescient on this point. Evolution, like so many other scientific disciplines, has indeed become useful.

In Chapter 2, I traced the origins and history for select domesticated animals and plants; dogs from wolves, work horses from wild horses, cattle from extinct wild aurochs, chickens from red junglefowl, bread wheat from wild grasses, Brussels sprouts from wild cabbage, and coffee from wild *Coffea arabica.* These and other domestic or cultivated species provide our primary sources of food, clothing, and shelter. Their central role in human history, including their great economic value to past and present human societies, demonstrates that artificial selection has been a key innovation for our species. Domestication via artificial selection facilitated agriculture, which gave rise to production of surplus food and food storage. This, in turn, allowed population sizes to grow, made more time available for pursuits beyond subsistence, and enhanced further development of culture.

The practice of artificial selection, beginning over 12,000 years ago, is based on promoting the reproductive success of individuals with desired traits over those with less-desired traits to enhance the future utility of succeeding generations of domesticated species. This required a common-sense understanding of the heritability of traits, a keen eye for discerning beneficial variation among populations of individuals, and a concept of creating change in organisms over time. Humans, by intentionally exerting the same sort of evolutionary pressure on populations of organisms as exerted in natural environments on wild populations, have taken evolution into their own hands. They intended simply to improve or maintain traits rather than study abstract evolutionary principles. Nevertheless, the results, gained through experience, trial and error, and teaching across generations, have transformed human existence. Artificial selection has resulted in rapid evolutionary change in organismal features of morphology, reproductive output, reproduction timing, physiological tolerances, nutritive value, flavor, and more. The processes involved are the same as those that have been at work in generating the broad diversity of all life forms.

Now that we have integrated our practical knowledge of artificial selection with a theoretical understanding of other evolutionary issues such as population variation, natural selection, and organismal diversification (including the origins of new and distinctive taxa), we can see the bigger picture of change over much longer time periods, and we have better understanding of the mechanisms of evolution. This understanding derives from research conducted at the level of genomes and cells as well as that of individuals and populations of organisms. We have watched not only successes of artificial selection but failures as well. We can, if we choose, apply our understanding of evolutionary processes to fix or avoid known problems. In the context of domestication of species, we understand the perils of excessive inbreeding, leading over time to increased incidence of unhealthy traits. We understand, as a corollary, the benefits of outbreeding, which enhances genetic diversity within populations. We can apply our knowledge of phylogeny and the tree of life to identify wild species and populations harboring the genetic variation that is best suited to boosting genetic diversity in various overworked domestic lineages. We can also apply phylogeny to identify candidates for potential development as cultivated species and artificial selection. Certain types of work will become increasingly important, such as studies of close relatives to species with known desired traits, but with naturally evolved adaptations to survive in particular environments and to withstand particular parasites or pathogens. Genetic engineering of domesticated plant and animal taxa to alter specific genes and monitor their effects is a logical extension of human efforts over thousands of years in evolving taxa for greater utility to people. The intent is the same, though the technology and awareness of evolutionary mechanisms is radically different. As we learn the function of particular genes in particular taxa, we can experiment to see how individual organisms succeed with those genes enhanced, blocked, or transplanted into individuals from species or breeds lacking them. Applying artificial selection via genetic engineering wisely rather than foolishly will be challenging. It seems inevitable that we will try, however, and that we will learn a great deal more about what can be evolved.

In Chapter 3, I discussed ways in which evolution and evolutionary thinking are applied to public health and medicine. Evolutionary theory provides the necessary context for understanding the nature of epidemic diseases. Infectious pathogens and human immune systems engage in continual coevolution: potentially adaptive changes in pathogens elicit adaptive responses in their hosts, which, in turn, elicit further change in pathogens. The virulence of pathogens and our resistance to them evolve in complex ways that are not well understood, but by taking an evolutionary approach, we can at least understand their change over time and determine measures we might take to influence that change, potentially reducing transmission rates and virulence in populations. Evolutionary population geneticists working in this field have the complex task of identifying relevant sets of genes, quantifying their variation in populations of both pathogens and hosts, and using those data to study evolutionary and ecological processes, with the goal of influencing them in ways that promote human health.

Knowledge of the basic features of natural selection is applied as artificial selection in development of attenuated and genetically engineered vaccines. Selection theory is also applied in an approach known as "directed evolution of molecules," focused on evolving molecules with enhanced traits to be used as medicines, antibodies, antigens, environmentally friendly solvents, and agents capable of binding to and protecting specific target molecules.

Accurate phylogenies for virus, bacteria, and eukaryotic pathogens help scientists identify the origins of human pathogens and, in many instances, their natural, nonhuman hosts. Understanding the evolutionary origins for human pathogens is useful in identifying basic precautions against future infection. Because functionally important traits tend to be shared by close relatives, understanding phylogenetic origins leads to reasoned working hypotheses about the life histories, transmission mechanisms, and host range for new pathogens. Evolutionary methods can potentially assist in identifying infectious pathogens associated with chronic rather than acute disease as well. Our susceptibility to diseases and various physical problems can be explained, thanks to evolutionary theory, as a legacy of our history of descent with modification from other animals. An evolutionary view of public health and medicine considers organisms in their environments as the unfinished result of millions of years of chance events and adaptive change. This view informs improved public health policy and health care in general.

Chapter 4 provided an overview of the earth's diversity of organisms and their lifestyles, followed by a discussion of the value of that diversity to people. Traditionally, value is recognized in terms of products derived from wild species. That value is substantial, and includes products such as food, heating and housing materials, feed for domestic animals, fiber for clothing, drugs and drug compounds, and enzymes for industry. More recently, we have begun to understand the role and value of biological diversity in the maintenance of the functional ecosystems on which human societies depend. This role includes protection of water and soil resources, regulation of climate, and the cycling of nutrients and waste products.

A basic task in any effort to manage and conserve resources, including biodiversity, is taking inventory. This entails surveying resident organisms from all of the earth's environments and analyzing them to understand their variety, life histories, distributions, and functional roles within ecosystems. Phylogenetic analyses are integral to the basic task of inventorying and classifying biological diversity for organismal lineages and their genes. Phylogenetics is also applied in prioritizing targets for conservation efforts seeking to maximize organismal and genetic diversity, in searching for potentially useful natural compounds (bioprospecting), and in enforcing conservation regulations. Because of our increased understanding of evolutionary processes, including the consequences of variable life-history strategies and reductions in population size, we are better able to maintain viability and genetic diversity within taxa that are endangered or at risk.

In Chapter 5 I examined how an evolutionary approach can provide insight into the history of cultural change, focusing on languages and religions. Language is our foremost window to the human mind, allowing us to describe our thoughts, plans, and emotions. Religions embody our ethical traditions and our views on the meaning of human lives. Understanding the history of change in the varieties of language and religion can help us understand ourselves and see the many cultural differences as variations on a theme rather than barriers between unrelated foreign traditions.

Phylogenetic methods developed by evolutionary biologists for reconstructing historical change among organisms may be applied with caution to estimate historical relationships among languages, where suitable data on cognate words (having the same root or origin) exists. Though the mechanisms of change differ between biological and cultural evolution, some analytical methods, including phylogenetic reconstruction, can be shared in assessing their respective evolutionary histories. Depicting historical relationships for entire cultural traditions and their features makes it possible, in some cases, to identify which traits are shared because of common cultural origins, which are shared because of cultural borrowing, and which have arisen independently on multiple occasions. Because heritability of cultural features is not genetically encoded, and adoption or borrowing of features between disparate traditions is common, the exact path of historical changes can be impossible to determine in some instances. However, there are some religious cultural claims, involving origins of peoples or relationships among peoples, that can be tested with modern molecular methods. Evolutionary methods can be applied to enhance our understanding of ourselves.

In Chapter 6, I discussed the changing role of evolutionary biology over time in two cultural arenas in the United States, the legal system and the public schools. I make the case that within the legal system as a whole, the role of evolution has expanded from that of an unappreciated science, worthy of suppression and susceptible to misuse by the courts. Evolutionary analyses are now better appreciated and widely used in criminal trials, linking (or de-linking) criminal evidence with suspects. Forensic analyses using DNA are increasingly common. Evolution predicts that some rapidly changing characters, such as hypervariable DNA sequences, will be unique to all individuals (excepting identical twins), and that more closely related individuals will share more recently derived characters than will distant relatives. Another way to state this is that DNA forensics is based on the identification and comparison of homologous genes and traits, shared due to common descent, in different individuals. The only way to discriminate between homologous and convergent similarity is in the context of evolution, specifically phylogenetic analyses. If the notion of homology for shared traits resulting from common descent is invalid, forensic DNA analyses lose their basic rationale. Variation in life history traits for various insects and other organisms associated with crime-scene evidence is evolutionary in its origin and has been employed successfully in an increasing number of court cases.

From the standpoint of science teachers, evolutionary biology serves as the unifying concept for the life sciences. Evolutionary history is the common crucible through which all life forms have passed, and by scientific investigation of evolutionary processes, students (and researchers) can better understand why and how any biological system works. This means that evolutionary biology, as a set of concepts and methods, is necessary for comprehensive biological understanding.

Evolution: Social Acceptance and Prognosis

Evolution provides the necessary context for understanding origins, functions, and maintenance of all biological systems, from genomes, to cells, to individuals and populations. For this reason, understanding the 3.8-billion-year history of life's evolution and figuring out its primary mechanisms of change stand as one of the great intellectual achievements of humans. Developing testable hypotheses as well as the data and analytical methods to discover the history of life's diversification ranks right up there with discoveries of electricity and the structure of matter. Those who studied physics and astronomy in its early days made observations repeatedly from a single location, but discovery of our evolutionary history required some observation of variation in organisms and comparisons over large geographic areas. In this light, it is not surprising that many basic discoveries in the physical sciences preceded those in biology.

The many applications of evolution provide additional proof for the basic idea, regardless of whether it is applied consciously or unconsciously. It is not necessary for an individual to understand details of mechanism to see that something works. No one needs to understand details of electromagnetism to turn on a light or shine it on the page of a book. Similarly, intentional artificial selection is applied evolution. If artificial selection is explained by practitioners as simple common sense rather than evolutionary biology, it makes no difference in terms of utility. After all, selection as differential reproductive success among individuals resulting in change over time is a simple truism; it just happens, and requires no technology unavailable to early humans.

Utility and repeated applications further strengthen the theory of evolution as an important component of modern science. Evolutionary theory, that is, the body of scientific concepts, facts, and analytical methods that explain evolution, contributes to our understanding of the origins and assembly of the various parts of cells of organisms. Our knowledge that mitochondria and chloroplasts are derived from bacterial ancestors is based in part on phylogenetic analyses of their genomes. If evolutionary theory were discredited or excised from our consciousness somehow, there would be no rational, compelling explanation for development of cells. Discrediting evolutionary theory would require discrediting much scientific research, as well as the scientific methods of observation and hypothesis testing on which the research is based. Something as basic as hypothesis testing cannot be deemed valid in some realms of science, such as physics, but invalid in others, such as biological history or population genetics. Selective application of the scientific method is not scientific. Thus, the successful repeated use of evolutionary theory should help in demonstrating, even publicizing, its inseparability from overall scientific theory.[2]

There is a lengthy argument made by some creationists and intelligent-design proponents that their disagreements with biologists boil down to conflicting philosophical positions. They would cast biological science and reason as mere philosophical options that favor natural causation over supernatural causation (naturalism versus theistic realism).[3] They then imply that these different philosophical positions have roughly equal empirical justification, with science weakened by its inability to discern what are assumed to be the real supernatural powers (despite their being undetectable by definition). It is also implied that the philosophical positions are mutually exclusive, and that belief in one or the other is a matter of personal choice. Because creationists and intelligent-design advocates believe such issues to be a matter of philosophical choice, both are considered to have equal claim to be taught in science classes. This view ignores the illogic of promoting belief in the supernatural as a valid scientific method or option. Scientific theories can be declared invalid if evidence contradicts them, but intelligent-design proponents' and creationists' views are based strictly on faith and cannot be tested with evidence.

Growing recognition of the applications and utility of evolution may well increase the general public's understanding and appreciation for evolution, however. When a concept and its resulting applications become useful, people tend to embrace the applications and, eventually, the underlying concepts. It is difficult to argue with success. Some may still choose not to believe in evolution, but the choice entails denial of more and more real-world demonstrations and uses of evolutionary biology.

Why has general acceptance of evolution taken such a long time? In Chapter 1, I traced the histories and time required for three unpopular discoveries to progress from their first heretical proposals to general acceptance, including acceptance by religious leaders, approximately 300 years later. I pointed to Pope John Paul II's 1996 statement that evolution was "more than a hypothesis" as one indication of acceptance by mainstream religions, and there are clergy within all the primary religious traditions that accept the best estimates of science regarding the antiquity of the earth and common descent for all species. Obviously, many who place less stock in evidence say their religious beliefs contradict and outweigh an evolutionary view of species relationships and earth history, and they tend to be more outspoken.

There has been no scientific alternative to evolution as the explanation for life's origins, history, and mechanisms of change since Darwin's time. The study of how evolution has proceeded, not if it has, is one of the most vibrant areas of modern scientific research, animated by exciting controversies about the mechanisms and timing of change, and fueled to a large extent by the expanding database of gene and whole genome sequences for diverse organisms. As just one indication of the growing influence of evolution within the life sciences, David Hillis showed that the number of peer-reviewed publications presenting or making use of phylogenetic trees has increased exponentially over the past two decades, approaching 5,000 annually by 2001. He noted, "With phylogeny as a framework, molecular biology could move from a largely descriptive science to a field of explanation and prediction."[4] The difference between having a DNA sequence for a gene from one species and having the same gene sequence from thousands of species, accompanied by their phylogenetic tree, vastly expands scientists' ability to explain and predict the causes and mechanisms of change over time at the molecular level, linking them with the unique lifestyles that those molecular changes facilitate in different species.

Creationists' assertions that evolutionary science is wrong all share an explicit agenda to advance religious views. The absence of nonfundamentalist proponents making their case is telling. And the acknowledged motivation of most creationist and intelligent-design proponents is to promote religious views in science classes. William Dembski, an advocate of intelligent-design and creationist views, states, "What drives me in this is that I think God's glory is being robbed by these naturalistic approaches to biological evolution . . . I want to see God get the credit."[5] Stipulating the outcome of an investigation of natural phenomena is also inherently unscientific.

The activities of creationists and intelligent-design proponents are understood to be nonscientific, both by scientists and by the courts when and where the issue has been considered. Scientists are entirely justified in viewing the proponents as either naïve or disingenuous in claiming their concerns about evolution to be scientific. The protests that "both sides of the debate" about evolution should be heard in science classes have little or no support outside of creationist circles, because the debate is not actually about the science. Creationists intend to appeal to Americans' sense of democracy and fair play. However, creationism has been presented within the scientific arena and it has been rejected. Losing in a fairly contested scientific dispute does not entitle the losing party to continual hearings; a flawed argument, if unchanged, does not improve with time.

Nor is scientific accuracy determined by public opinion. Ideally, scientists seek to identify and test potential facts regardless of the sentiments of the public or special interest groups. Science is democratic in the sense that participation is open, and consensus based on independent analyses is valued. This does not mean that truth resides somewhere between the extremes, or that facts should conform to heartfelt conviction. Mass protest does not change the fact that the earth orbits the sun or that man is a primate. Creationists who attempt to refute evolution and weaken science education by promoting supernatural causation as an explanation for complex natural phenomena are in dubious historical company. They stand alongside those who resisted and denied other unpopular discoveries, long after numerous, independent lines of evidence showed them to be mistaken.

Despite acceptance of evolution by the scientific community and the broad availability of the evidence and analyses in its support, resistance to evolution within some religious communities appears certain to continue. Is evolution different from other discoveries at odds with religious traditions? Arguably, yes, evolution is different in the immediacy of its perceived threat. It is perceived as a direct threat to the role of God in the material origins of humans. And explanation of human origins, whether for our species or for each individual, is about as personal as it gets. Some of the ire of creationists is focused on scientific reasoning in general; however, defending the version of human origins presented in Genesis remains a primary concern for them. Many religious leaders are now satisfied to consider humans as evolving naturally from other life forms at the instigation of God. This entails a figurative rather than literal interpretation of Genesis, also necessary for acceptance of the earth's orbiting of the sun. There is an unfortunate and mistaken notion abroad that accepting the fact of evolution entails a life without morals or spirituality. The primary values in science—open inquiry, truth-telling, and personal integrity—are worthy of our best institutions. The history of human affairs, particularly in the case of wars or massacres conducted in the name of God, shows there is no necessary connection between professing belief in God and the practice of ethical behavior.

The controversy between creationists and others is really about cultural dominance. It involves conflicting ideas about the methods for gaining knowledge of the world and about claims of moral authority. In *Culture Wars,* James Hunter writes, "Whatever else may be involved, cultural conflict is about power—a struggle to achieve or maintain the power to define reality." In a similar vein, Kary Smout notes, "This controversy is ultimately a debate about who gets to clothe virtue in the language of truth."[6] Given the high stakes and the disparate natures of scientific inquiry and faith, there is no reason to think the contest for cultural ascendancy will disappear anytime soon. Those who care about the health of the scientific enterprise will ask how science education, including an understanding of evolutionary biology, can be strengthened and kept free from religious or ideological restrictions in multicultural societies.

If there is to be any commerce between science and religion, then, on what footing can the discussion commence? Four general categories of interaction between science and religion have been outlined by Ian Barbour: integration, dialogue, conflict, and independence.[7] All have been pursued to one degree or another, and preference for one approach over another is largely a matter of personality and ideology. Let's briefly consider how these variant approaches have affected public views on evolution and creation.

Integration entails reformulations of orthodox religious beliefs. For example, one might reformulate a literal interpretation of the biblical account of Genesis to include the view that God works through natural processes to create life, including the process of evolution, which would require billions of years rather than six days. Reformulations of religious orthodoxy are as old as religion itself. There is not a single religion that is the same today as it was in its historical beginnings. Thomas Aquinas and Martin Luther are two reformers who worked from within Christian religious tradition as members of the clergy. Darwin may be considered a reformer of sorts, though from the outside. He defended, at least on one occasion, the idea that God may have initiated the laws of the evolutionary process. The causes for reformulation are many. Reason, the basic approach of science, seeks explanations consistent with observed reality and has been the primary cause for reformulation over the past thousand years or so. In this sense, integration for science and religion can have a positive effect on public appreciation of science and evolution. Of course, attempted integration can have the opposite effect when pseudo-sciences like intelligent design are recruited in support of supernatural phenomena.

Dialogue, according to Barbour, arises from consideration of similarities between science and religion. Similarities can be seen in the presence (but not the specifics) of underlying assumptions and some analogous methods and concepts. It is not clear what can be drawn from these analogies, however. Minimally, dialogue can improve understanding of the limitations of both science and religion, although the nature of their limitations are quite different.

Conflict accentuates inherent differences in science and religion. It can be seen in many of the public debates about creationism or intelligent design versus evolution. These are variously good theater, bad theater, and simple invective. They have considerable power to provoke the opposition and rally the supporting troops. Creationists and intelligent-design proponents often feel they win merely by showing up and forcing scientists to explain the principles of evolution, especially if this takes place in an academic or scientific setting. Their argument for God, they believe, is being heard in the home of excessive empiricism. Scientists feel they can win simply by exposing the creationists as scientific pretenders and devotees of the irrational. Few minds are changed. Because creationists attempt to cast doubt on evolution as science, the scientific community must respond forcefully, especially when important political issues arise. Conversely, scientists have spent little time casting doubt on creationists and intelligent-design proponents as representatives of the very diverse religious community, though doing so could be effective from a political standpoint. Creationists and intelligent-design proponents do not speak for the large religious population that has no quibble with modern evolutionary biology and reconciles evolution with a personal belief in God. A 2001 Gallup poll found that 37 percent of Americans believe that "man developed over millions of years from less advanced forms of life but God guided the process."[8] Creationists invoke literal interpretation of biblical passages, but even within their own ranks there is great inconsistency in exactly what the literal interpretation is, and whether or not it need be embraced. Intelligent-design proponents, including many creationists, often steer discussion away from their particular interpretations of Genesis, largely because arguments in favor of teaching a literal, scientific accuracy for Genesis in science classes do not fare well in court.

Science classes are seen by some religious activists as the key arena for the conflict they want to engage. Young and impressionable minds are exposed to widely respected scientific methods and views for explaining the natural world. The difficulty for the activists is that by demanding consideration for their views within the scientific arena, they invite the scrutiny that exposes their views as nonscientific. Even worse for the activists, if faith-based assertions about God or an unnamed intelligent designer were examined in the same manner as all other science-class topics, that is as testable hypotheses, the test results would inevitably be negative. Few religious groups would be so confused, but seeking validation of religious tenets from science suggests that science might refute them as well.

Independence as a category places science and religion in separate domains, and is motivated by a desire to reduce conflict and respect the distinctive characters of science and religion. Gould's view of nonoverlapping magisteria presents an example of this noninteraction. Independence of science and religion displeases those inclined toward conflict, by its accommodation of the sensitivities of others and by restricting those who feel justified, if not compelled, to pass judgment on the core questions claimed by others. I agree with the premise of mutual respect; however, this approach is ultimately unsatisfactory because the separation is artificial. For example, both science and religion would like to treat the origin and maintenance of morality. Have morals arisen as a mandate from God, who rewards compliance and punishes transgressors? Or have they evolved as a natural, potentially adaptive social contract among individuals living in groups, with compliance tending to enhance reproductive success of individuals and their relatives? Religious faith and scientific inquiry cannot both be right when they reach different explanations for the same phenomena. Gould's solution, favored by Galileo too, is to place moral discourse and the supernatural within the religious domain and the natural, tangible world within that of science. However, many religionists would not willingly turn over the capacity for explaining all aspects of the natural world to science. Correspondingly, those who do not draw their views on morality from religious authority or sacred documents will not forgo their own rationale and scientific evidence in support of natural origins for ethical behavior. Should an evolutionary anthropologist ignore evidence of apparent altruistic behavior among a certain species of primate because it has been agreed that the subject falls in the religious domain? The fact that the Golden Rule has been embraced as an ethical ideal by many more people and cultures than has any particular religion suggests that its motivations are not necessarily religious. The forced separation and arbitrary walls (wherever placed) prevent the independence approach from providing an ultimate resolution for many of the points of conflict.

Greater independence for science and religion can be found, however, in their alternative primary pursuits of facts by science and of meaning (or values) by religion, and their alternative criteria of evidence and faith. If the independence of scientific and religious domains were based on the distinctive intentions of the individuals making claims and the criteria used, instead of the topics addressed and criteria claimed, the independence category of relationship would be improved. In that situation, discourse on the meaning of human lives from nonreligious scholars would be assessed as such, and the views about science put forth by creationists and intelligent-design proponents would be evaluated as the faith-based assertions they are. Independence fails as a final resolution to the conflicting claims of science and religion, but as an interim agreement this view has the appeal of truth in advertising for the works of scholars and demagogues alike.

Loyal Rue makes a compelling case for the biological and social origins of religion.[9] He discusses religions as human traditions that seek to influence our behavior and culture in ways that benefit both individuals and groups. This is difficult because behaviors that might benefit an individual, such as theft or deceit, may well be detrimental for groups as a whole. Cultural traditions that link self-esteem and rewards for individuals to behaviors more conducive to successful human interactions are important for functional societies. This provides a natural theory of religion, explaining both its positive and negative influences on human welfare, and this approach is likely to become increasingly important in discussing the roles of science and religion.

Humans have a deep-seated desire to explain the physical world. If every shred of human culture, from sacred documents to the protocols for sequencing genomes, were somehow eradicated, our species would likely start again with supernatural beliefs, which would provide immediate answers for complex natural phenomena, and we would slowly work out the natural explanations as observational tools and empirical analyses improved. Supernatural beliefs would again become traditions and convictions, helping to define social groups, and natural explanations would meet resistance for being difficult to grasp, at best, and heresy, at worst.

I suspect that people will continue to do as they have done in the past—adopting a variety of positions on evolution depending largely on their cultural environment. Despite evolution's acceptance by mainstream science and the reliable utility of its concepts and methods, resistance to evolution is likely to continue within some groups, in part because reaping the benefits of science does not require any understanding of the underlying scientific concepts and methods. Jacob Bronowski wrote of this problem: "The body of technical science burdens and threatens us because we are trying to employ the body without the spirit; we are trying to buy the corpse of science."[10] People have great appetite for the products and services of science, but often little appreciation of the scientific worldview giving rise to them. After successful medical treatments, it is not uncommon for grateful patients to be effusive in crediting divine intervention, neglecting to acknowledge the many researchers that discovered the principles involved in the treatments and the crucial role of the scientific method in improving their lives.

It will remain an ongoing challenge for educators to communicate the values, excitement, and significance of science, in addition to its technical details, to successive generations of students. It will also be their challenge to reduce the perceived threat that evolution presents to some religious groups. There is a role for the "take-no-prisoners" approach in exposing the pseudo-science and false claims of creationists and intelligent-design activists. However, religious faith continues to provide great comfort to many, and a tolerant and pragmatic approach, focused on opening the minds of students and the public to the scientific enterprise and including the ability to think critically, may accomplish much more.

To succeed within religious cultures, evolution needs to be decoupled from people's fears that descent with modification for all species somehow severs their link to religion, morality, or God. Evolution's opponents do not want that to happen, but the long-term cultural trend away from fear is already established. Many people with strong religious convictions accept the findings of science regarding evolution. They are not afraid that God's glory is being robbed by science. Far from it. In the spirit of integration, they may believe that a divine power initiated the evolutionary process. Integration and reformulation of this kind has a history as long as that of human culture. It is demonstrated in the acceptance of heliocentrism and germ theory, which were at one time deemed heretical but are now widely accepted, even by fundamentalists. Many devout believers in God reconcile the earth's orbit around the sun and other scientific facts with a figurative, rather than literal, interpretation of the biblical passages that appear to contradict them. Many, including many clergy, already do the same for evolution, and it seems likely that the practical value and repeated applications of evolutionary biology will continue to influence religious cultural change. Cultural evolution and change over time for religious traditions as described in Chapter 5 are undeniable, and as fascinating and complex as the evolution of organisms. Some squirm as the worm turns and religious traditions are called to account. However, understanding of evolution can illuminate the human motivation for the ethical behavior that our religious traditions seek to promote. In this sense, an evolutionary view can be recognized as a potential ally, not a threat. After all, wouldn't it be a weak faith in God that fears rather than welcomes scientific discoveries? The long history of religion accommodating once heretical discoveries about the natural world suggests that the fact of evolution will, eventually, complete its cultural rounds.

I began this chapter with a story about Benjamin Franklin watching the first manned flight in a hot-air balloon. At one time, evolutionary biology was like Franklin's metaphorical newborn baby. It was a novel conception of the origins of humans and of all life, rooted firmly in observations of nature and testable mechanisms of heritable change across generations. As Franklin might have predicted, learning as much as possible about the nature of organic change over time, learning how to infer past evolution, and how to predict and influence future evolution have indeed become useful. As with any science, the knowledge gained can be used for good, bad, or indifferent purposes; it is our responsibility to use the knowledge wisely.

Evolution as common cause explains many seemingly disconnected observations about nature. Phenomena not realized to be evolutionary at the time of discovery can now be understood and linked to each other in their explanations. This includes understanding how to domesticate wild species for agriculture; how to manage our exposure to protist, bacterial, and viral pathogens; how to manage our environment to help maintain the diversity of species and functioning of ecosystems that are vital to human lives; how humans represent a single evolutionary family with variant cultures but shared biological capabilities and motivations; and, last but not least, how to use knowledge of evolution in pursuit of justice within the legal system and in pursuit of further scientific discovery via education and academic research.

Evolution is not just useful, it has become indispensable.

NOTES

1. A Brief History of Three Unpopular Discoveries

1. George Rosen, *A History of Public Health* (Baltimore: Johns Hopkins University Press, 1958).
2. Geoffrey Wigoder, ed., *Encyclopedia Judaica* (Jerusalem: Keter Publishing House, 1996).
3. For a recent translation of Genesis seeking to identify sources, see Stephen Mitchell, *Genesis: A New Translation of the Classical Biblical Stories* (New York: Harper-Collins, 1996); Melvin Konner, *Unsettled: An Anthropology of the Jews* (New York: Viking Compass, 2003).
4. François Marie Arouet de Voltaire, 1778, "On Francis Bacon. Letters on the English or Lettres Philosophiques," in *French and English philosophers: Descartes, Rousseau, Voltaire, Hobbes: with Introductions and Notes* (New York: P. F. Collier, 1910, The Harvard Classics v. 34).
5. Marquis de Condorcet, *Sketch for a Historical Picture of the Progress of the Human Mind,* ed. Stuart Hampshire, trans. June Barraclough (London: Weidenfeld & Nicolson, 1955), p. 72; David Lindberg, *The Beginnings of Western Science* (Chicago: University of Chicago Press, 1992); Edward Grant, *The Foundations of Modern Science in the Middle Ages* (Cambridge: Cambridge University Press, 1996).

6. Lindberg, *Beginnings of Western Science.*
7. Lynn Thorndike, *University Records and Life in the Middle Ages* (New York: Columbia University Press, 1944); Fernand Van Steenberghen, *Aristotle in the West,* trans. Leonard Johnston (Louvain: Nauwelaerts, 1955).
8. See Edward Grant, *A Source Book in Medieval Science* (Cambridge, MA: Harvard University Press, 1974) for translations from Latin text edited by Heinrich Denifle and Emile Chatelain, 1889–1897, *Chartularium Universitatis Parisiensis,* 4 vols. (Fratrum Delalain, Paris), 1:543–555.
9. Casey Wood and F. Marjorie Fyfe, *The Art of Falconry: Being the "De Arte Venandi cum Avibus" of Frederick II of Hohenstaufen* (Stanford: Stanford University Press, 1943), p. xxxix.
10. Wood and Fyfe, *Art of Falconry,* p. 49.
11. Francis Bacon, *Novum Organum,* 1620, in *Great Books of the Western World,* ed. Robert Hutchins (Chicago: Encyclopedia Britannica, 1952), vol. 30, p. 111.
12. Francis Bacon, *New Atlantis,* 1624, in *Great Books of the Western World,* ed. Robert Hutchins (Chicago: Encyclopedia Britannica, 1952), vol. 30, p. 210.
13. Benoit de Maillet, *Telliamed: Or, Discourses Between an Indian Philosopher and a French Missionary, on the Diminution of the Sea, the Formation of the Earth, the Origin of Men and Animals, And other Curious Subjects, relating to Natural History and Philosophy* (London: T. Osborne, 1748), trans. and ed. Albert V. Carozzi (Urbana: University of Illinois Press, 1968), p. 188.
14. See Peter Bowler, *Evolution: The History of an Idea* (Berkeley: University of California Press, 1989), pp. 33–34.
15. Jean Baptiste Lamarck, 1809, *Zoological Philosophy,* trans. Hugh Elliot (Chicago: University of Chicago Press, 1984), pp. 56, 112.
16. Jean Baptiste Lamarck, 1815–1822, *Histoire Naturelle des Animaux sans Vertebres,* 6 vols. (Paris, rpt. Brussels: Culture et Civilisation, 1969).
17. Pietro Corsi, *The Age of Lamarck: Evolutionary Theories in France, 1790–1830* (Berkeley: University of California Press, 1988), p. xi; Richard W. Burkhardt, Jr., *The Spirit of System* (Cambridge, MA: Harvard University Press, 1995).
18. For discussion of Darwin's seminal contributions and their historical context, see Ernst Mayr, *The Growth of Biological Thought* (Cambridge, MA: Harvard University Press, 1982); Bowler, *Evolution: The History of an Idea;* Edward Larson, *Evolution: The Remarkable History of a Scientific Theory* (New York: Random House, 2004).

19. Francis Darwin, ed., 1887. *The Life and Letters of Charles Darwin.* 3 vols. (New York: Johnson Reprint Corporation, 1969).
20. Michael Shermer, *In Darwin's Shadow: The Life and Science of Alfred Russel Wallace* (New York: Oxford University Press, 2002).
21. Joel Cracraft and Michael J. Donoghue, eds., *Assembling the Tree of Life* (New York: Oxford University Press, 2004).
22. John Paul II, "The Pope's message on evolution and four commentaries," *Quarterly Review of Biology* 72 (1997): 381–406.

2. Domestication: Evolution in Human Hands

1. For comprehensive treatments, see Jared Diamond, *Guns, Germs, and Steel* (New York: Norton, 1997); Bruce D. Smith, *The Emergence of Agriculture* (New York: Scientific American Library, 1998); Juliet Clutton-Brock, *A Natural History of Domesticated Mammals* (Cambridge: Cambridge University Press, 1999); Jared Diamond, "Evolution, consequences and future of plant and animal domestication," *Nature* 418 (2002): 700–707.
2. B. Hare, M. Brown, C. Williamson, and M. Tomasello, "The domestication of social cognition in dogs," *Science* 298 (2002): 1634–1636.
3. D. K. Belyaev, "Domestication of animals," *Science Journal (U.K.)* 5 (1969): 47–52; D. K. Belyaev, "Destabilizing selection as a factor in domestication," *The Journal of Heredity* 70 (1979): 301–308; L. N. Trut, "Early canid domestication: The farm-fox experiment," *Am. Scientist* 87 (1999): 160–169.
4. Peter Savolainen, Ya-Ping Zhang, Jing Luo, Joakim Lundeberg, and Thomas Leitner, "Genetic evidence for an east Asian origin of domestic dogs," *Science* 298 (2002): 1610–1613; Jennifer A. Leonard, Robert K. Wayne, J. Wheeler, Raùl Valadez, Sonia Cuillén, and Carles Vilà. "Ancient DNA evidence for Old World origin of New World dogs," *Science* 298 (2002): 1613–1616.
5. R. Harcourt, "The dog in prehistoric and early historic Britain," *J. Archaeological Science* 1 (1974): 151–175.
6. For an overview of dog breeds, see Bruce Fogle, *The New Encyclopedia of the Dog* (London: Dorling Kindersley, 2000).
7. C. Vilà, J. A. Leonard, A. Gotherstrom, S. Marklund, K. Sandberg, K. Liden, R. K. Wayne, and H. Ellegren, "Widespread origins of domestic horse lineages," *Science* 291 (2001): 474–477; T. Jansen, Peter Forster, Marsha A. Levine, Hardy Oelke, Matthew Hurles, Colin Renfrew, Jürgen Weber, and Klaus Olek. "Mitochondrial DNA and

the origins of the domestic horse," *Proc. Nat. Acad. Sci. USA* 99 (2002): 10905–10910.

8. Janusz Piekalkiewicz, *The Cavalry of World War II* (London: Orbis Publishing Limited, 1979).
9. R. R. Loftus, D. E. MacHugh, D. G. Bradley, P. M. Sharp, and P. Cunningham, "Evidence for two independent domestications of cattle," *Proc. Nat. Acad. Sci. USA* 91 (1994): 2757–2761.
10. Page Smith and Charles Daniel, *The Chicken Book* (Athens: University of Georgia Press, 2000).
11. Attributed to the written works of Diodorus Siculus, a first century BCE Greek historian, in Thomas Browne, *Pseudodoxia Epidemica* (London: Printed by J. R. for Nath. Ekins, 1646).
12. Ian L. Mason, ed., *Evolution of Domesticated Animals* (London: Longman, 1984).
13. Daniel Zohary and Maria Hopf, *Domestication of Plants in the Old World* (New York: Oxford University Press, 2000).
14. Mark Pendergrast, *Uncommon Grounds* (New York: Basic Books, 1999).
15. Lashermes, P., M.-C. Combes, J. Robert, P. Trouslot, A. D'Hont, F. Anthony, and A. Charrier, "Molecular characterisation and origin of the *Coffea arabica* genome," Mol. Gen. Genet. 261 (1999): 259–266.
16. M. B. Silvarolla, P. Mazzafera, and L. C. Fazuoli, "Plant biochemistry: A naturally decaffeinated arabica coffee," *Nature* 429 (2004): 826.
17. Kenneth D. White, *Roman Farming* (Ithaca: Cornell University Press, 1970); Zohary and Hopf, *Domestication of Plants in the Old World.*
18. R. Brosch, S. V. Gordon, M. Marmiesse, P. Brodin, C. Buchrieser, K. Eiglmeier, T. Garnier, C. Gutierrez, G. Hewinson, K. Kremer, L. M. Parsons, A. S. Pym, S. Samper, D. van Soolingen, and S. T. Cole, "A new evolutionary scenario for the *Mycobacterium tuberculosis* complex," *Proc. Nat. Acad. Sci. USA* 99 (2002): 3684–3689.
19. Steve Jones, *Darwin's Ghost: The Origin of Species Updated* (New York: Random House, 1999).

3. Evolution in Public Health and Medicine

1. For detailed reviews of phylogenetic methods, see D. L. Swofford, G. J. Olsen, P. J. Waddell, and D. M. Hillis, "Phylogenetic inference," in D. M. Hillis, C. Moritz, and B. K. Mable, eds., *Molecular Systematics* (Sunderland: Sinauer Associates, 1996), pp. 407–514; Roderic D. M. Page and Edward C. Holmes, *Molecular Evolution: A Phylogenetic Ap-*

proach (Oxford: Blackwell Publishers, 1998); Joseph Felsenstein, *Inferring Phylogenies* (Sunderland: Sinauer Associates, 2004).

2. J. Hietpas, L. K. McMullan, D. P. Mindell, H. L. Hanson, and C. M. Rice, "Keeping track of viruses," in Roger G. Breeze, Bruce Budowle, and Steven E. Schutzer, eds., *Microbial Forensics* (Burlington, MA: Elsevier Academic Press, 2005), 55–97.
3. D. Baltimore, "Expression of animal virus genomes," *Bacteriological Reviews* 35 (1971): 235–241.
4. C. M. Fauquet, M. A. Mayo, J. Maniloff, U. Desselberger, and L. A. Ball, eds., *Virus Taxonomy* (San Diego: Elsevier, 2005).
5. R. S. Lanciotti, J. T. Roehrig, V. Deubel, J. Smith, M. Parker, K. Steele, B. Crise, K. E. Volpe, M. B. Crabtree, J. H. Scherret, R. A. Hall, J. S. MacKenzie, C. B. Cropp, B. Panigrahy, E. Ostlund, B. Schmitt, M. Malkinson, C. Banet, J. Weissman, N. Komar, H. M. Savage, W. Stone, T. McNamara, and D. J. Gubler, "Origin of the West Nile virus responsible for an outbreak of encephalitis in the northeastern United States," *Science* 286 (1999): 2333–2337; R. S. Lanciotti, G. D. Ebel, V. Deubel, A. J. Kerst, S. Murri, R. Meyer, M. Bowen, N. McKinney, W. E. Morrill, M. B. Crabtree, L. D. Kramer, and J. T. Roehrig, "Complete genome sequences and phylogenetic analysis of West Nile virus strains isolated from the United States, Europe, and the Middle East," *Virology* 298 (2002): 96–105.
6. J. F. Anderson, C. R. Vossbrinck, T. G. Andreadis, A. Iton, W. H. Beckwith, and D. R. Mayo, "A phylogenetic approach to following West Nile virus in Connecticut," *Proc. Nat. Acad. Sci. USA* 98 (2001): 12885–12889.
7. P. M. de A. Zanotto, E. A. Gould, G. F. Gao, P. H. Harvey, and E. C. Holmes, "Population dynamics of flaviviruses revealed by molecular phylogenies," *Proc. Nat. Acad. Sci. USA* 93 (1996): 548–553.
8. Stephen R. Palmer, E. J. L. Soulsby, and D. I. H. Simpson, *Zoonoses: Biology, Clinical Practice and Public Health Control* (New York: Oxford University Press, 1998); Martin Shakespeare, *Zoonoses* (London: Pharmaceutical Press, 2002).
9. J. M. Morvan, V. Deubel, P. Gounon, E. Nakoune, P. Barriere, S. Murri, O. Perpete, B. Selekon, D. Coudrier, A. Gautier-Hion, and M. Colyn, "Identification of Ebola virus sequences present as RNA or DNA in organs of terrestrial small mammals of the Central African Republic," *Microbes and Infection* 1 (1999): 1193–1201.
10. K. L. Desphande, V. A. Fried, M. Ando, and R. G. Webster, "Glycosylation affects cleavage of an H5N2 influenza virus hemagglutinin and regulates virulence," *Proc. Nat. Acad. Sci. USA* 84 (1987): 36–40; T.

Horimoto and Y. Kawaoka, "Pandemic threat posed by avian influenza A viruses," *Clinical Microbiology Reviews* 14 (2001): 129–149.

11. R. M. Bush, C. A. Bender, K. Subbarao, N. J. Cox, and W. M. Fitch, "Predicting the evolution of human influenza A," *Science* 286 (1999): 1921–1925.

12. P. M. Sharp, E. Bailes, R. R. Chaudhuri, C. M. Rodenburg, M. O. Santiago, and B. H. Hahn, "The origins of acquired immune deficiency syndrome viruses: Where and when?" *Philos. Trans. R Soc. Lond. B* 356 (2001): 867–876; E. C. Holmes, "On the origin and evolution of the human immunodeficiency virus (HIV)," *Biol. Rev.* 76 (2001): 239–254.

13. A. J. L. Brown, E. C. Holmes, "Evolutionary biology of human-immunodeficiency-virus," *Annu. Rev. Ecol. Syst.* 25 (1994): 127–165; D. P. Mindell, "Positive selection and rates of evolution in immunodeficiency viruses from humans and chimpanzees," *Proc. Nat. Acad. Sci. USA* 93 (1996): 3284–3288; B. Foley, H. Pan, S. Buchbinder, and E. L. Delwart, "Apparent founder effect during the early years of the San Francisco HIV type 1 epidemic (1978–1979)," *AIDS Res. Hum. Retroviruses* 16 (2000): 1463–1469.

14. T. Leitner, D. Escanilla, C. Franzen, M. Uhlen, and J. Albert, "Accurate reconstruction of a known HIV-1 transmission history by phylogenetic tree analysis," *Proc. Nat. Acad. Sci. USA* 93 (1996): 10864–10869.

15. L. Nguyen, D. J. Hu, K. Choopanya, S. Vanichseni, D. Kitayaporn, F. van Griensven, P. A. Mock, W. Kittikraisak, N. L. Young, T. D. Mastro, and S. Subbarao, "Genetic analysis of incident HIV-1 strains among injection drug users in Bangkok: Evidence for multiple transmission clusters during a period of high incidence," *J. Acquired Immune Deficiency Syndromes* 30 (2002): 248–256; C.-Y. Ou, C. A. Ciesielski, G. Myers, C. I. Bandea, C.-C. Luo, B. T. M. Korber, J. I. Mullins, G. Schochetman, R. L. Berkelman, A. N. Economou, J. J. Witte, L. J. Furman, G. A. Satten, K. A. MacInnes, J. W. Curran, H. W. Jaffe, L. I. Group, and E. I. Group, "Molecular epidemiology of HIV transmission in a dental practice," *Science* 256 (1992): 1165–1171; R. W. DeBry, L. G. Abele, S. H. Weiss, M. D. Hill, M. Bouzas, E. Lorenzo, F. Graebnitz, and L. Resnick, "Dental HIV transmission?" *Nature* 361 (1993): 691; D. M. Hillis and J. P. Huelsenbeck, "Support for dental HIV transmission," *Nature* 369 (1994): 24–25; K. A. Crandall, "Intraspecific phylogenetics: Support for dental transmission of human immunodeficiency virus," *J. Virol.* 69 (1995): 2351–2356; M. L. Metzker, D. P. Mindell, X.-M. Liu, R. G. Ptak, R. A. Gibbs, and D. M. Hillis, "Molecular evidence of HIV-1 transmission in a U.S. criminal case,"

Proc. Nat. Acad. Sci. USA 99 (2002): 14292–14297; C. P. Goujon, V. M. Schneider, J. Grofti, J. Montigny, V. Jeantils, P. Astagneau, W. Rozenbaum, F. Lot, C. Frocrain-Herchkovitch, N. Delphin, F. Le Gal, J. C. Nicolas, M. C. Milinkovitch, and P. Deny, "Phylogenetic analyses indicate an atypical nurse-to-patient transmission of human immunodeficiency virus type 1," *J. Virol.* 74 (2000): 2525–2532.

16. M. J. Gonzales, I. Belitskaya, K. M. Dupnik, S. Y. Rhee, and R. W. Shafer, "Protease and reverse transcriptase mutation patterns in HIV type 1 isolates from heavily treated persons: Comparison of isolates from Northern California with isolates from other regions," *AIDS Res. Hum. Retrov.* 19 (2003): 909–915.

17. M. Samson, F. Libert, B. J. Doranz, J. Rucker, C. Liesnard, C. M. Farber, S. Saragosti, C. Lapoumeroulie, J. Cognaux, C. Forceille, G. Muyldermans, C. Verhofstede, G. Burtonboy, M. Georges, T. Imai, S. Rana, Y. J. Yi, R. J. Smyth, R. G. Collman, R. W. Doms, G. Vassart, and M. Parmentier, "Resistance to HIV-1 infection in Caucasian individuals bearing mutant alleles of the CCR-5 chemokine receptor gene," *Nature* 382 (1996): 722–725; J. C. Stephens, D. E. Reich, D. B. Goldstein, H. D. Shin, M. W. Smith, M. Carrington, C. Winkler, G. A. Huttley, R. Allikmets, L. Schriml, B. Gerrard, M. Malasky, M. D. Ramos, S. Morlot, M. Tzetis, C. Oddoux, F. S. di Giovine, G. Nasioulas, D. Chandler, M. Aseev, M. Hanson, L. Kalaydjieva, D. Glavac, P. Gasparini, E. Kanavakis, M. Claustres, M. Kambouris, H. Ostrer, G. Duff, V. Baranov, H. Sibul, A. Metspalu, D. Goldman, N. Martin, D. Duffy, J. Schmidtke, X. Estivill, S. J. O'Brien, and M. Dean, "Dating the origin of the CCR5-Delta 32 AIDS-resistance allele by the coalescence of haplotypes," *Am. J. Hum. Genetics* 62 (1998): 1507–1515; S. R. Duncan, S. Scott, and C. J. Duncan, "Reappraisal of the historical selective pressures for the *CCR5-32* mutation," *J. Med. Genetics* 42 (2005): 205–208.

18. P. Keim, A. Kalif, J. Schupp, K. Hill, S. E. Travis, K. Richmond, D. M. Adair, M. Hugh-Jones, C. R. Kuske, and P. Jackson, "Molecular evolution and diversity in *Bacillus anthracis* as detected by amplified fragment length polymorphism markers," *J. Bacteriol.* 179 (1997): 818–824; P. Keim, L. B. Price, A. M. Klevytska, K. L. Smith, J. M. Schupp, R. Okinaka, P. J. Jackson, and M. E. Hugh-Jones, "Multiple-locus variable-number tandem repeat analysis reveals genetic relationships within *Bacillus anthracis,*" *J. Bacteriol.* 182 (2000): 2928–2936.

19. N. Ivanova, A. Sorokin, I. Anderson, N. Galleron, B. Candelon, V. Kapatral, A. Bhattacharyya, G. Reznik, N. Mikhailova, A. Lapidus, L. Chu, M. Mazur, E. Goltsman, N. Larsen, M. D'Souza, T. Walunas, Y. Grechkin, G. Pusch, R. Haselkorn, M. Fonstein, S. D. Ehrlich,

R. Overbeek, N. Kyrpides, "Genome sequence of *Bacillus cereus* and comparative analysis with *Bacillus anthracis*," *Nature* 423 (2003): 87–91; T. D. Read, S. N. Peterson, N. Tourasse, L. W. Baillie, I. T. Paulsen, K. E. Nelson, H. Tettelin, D. E. Fouts, J. A. Eisen, S. R. Gill, E. K. Holtzapple, O. A. Okstad, E. Helgason, J. Rilstone, M. Wu, J. F. Kolonay, M. J. Beanan, R. J. Dodson, L. M. Brinkac, M. Gwinn, R. T. DeBoy, R. Madpu, S. C. Daugherty, A. S. Durkin, D. H. Haft, W. C. Nelson, J. D. Peterson, M. Pop, H. M. Khouri, D. Radune, J. L. Benton, Y. Mahamoud, L. X. Jiang, I. R. Hance, J. F. Weidman, K. J. Berry, R. D. Plaut, A. M. Wolf, K. L. Watkins, W. C. Nierman, A. Hazen, R. Cline, C. Redmond, J. E. Thwaite, O. White, S. L. Salzberg, B. Thomason, A. M. Friedlander, T. M. Koehler, P. C. Hanna, A. B. Kolsto, and C. M. Fraser, "The genome sequence of *Bacillus anthracis Ames* and comparison to closely related bacteria," *Nature* 423 (2003): 81–86.

20. Paul W. Ewald, *Plague Time* (New York: Free Press, 2000).

21. J. S. Lawson, D. Tran, and W. D. Rawlinson, "From Bittner to Barr: A viral, diet and hormone breast cancer etiology hypothesis," *Breast Cancer Res.* 3 (2001): 81–85.

22. See C. Zimmer, "Do chronic diseases have an infectious root?" *Science* 293 (2001): 1974–1977.

23. P. W. Ewald, "Host-parasite relations, vectors, and the evolution of disease severity," *Annu. Rev. Ecol. Syst.* 14 (1983): 465–485; Ewald, *Plague Time.*

24. F. Macfarlane Burnet and Ellen Clark, *Influenza; a survey of the last 50 years in the light of modern work on the virus of epidemic influenza* (Melbourne: Macmillan and Co., 1942); Ewald, *Plague Time.*

25. D. Ebert and J. J. Bull, "Challenging the trade-off model for the evolution of virulence: Is virulence management feasible?" *Trends Microbiol.* 11 (2003): 15–20. A. P. Galvani, "Epidemiology meets evolutionary ecology," *Trends Ecol. & Evol.* 18 (2003): 132–139.

26. P. Williams, M. Camara, A. Hardman, S. Swift, D. Milton, V. J. Hope, K. Winzer, B. Middleton, D. I. Pritchard, and B. W. Bycroft, "Quorum sensing and the population-dependent control of virulence," *Philos. Trans. R. Soc. London Ser. B* 355 (2000): 667–680; B. J. Crespi, "The evolution of social behavior in microorganisms," *Trends Ecol. Evol.* 16 (2001): 178–183.

27. A. E. van den Bogaard and E. E. Stobberingh, "Epidemiology of resistance to antibiotics. Links between animals and humans," *Int. J. Antimicrob. Agents* 14 (2000): 327–335.

28. S. L. R. Kardia, M. B. Haviland, R. E. Ferrell, and C. F. Sing, "A search

for functional mutations in the lipoprotein lipase (LPL) gene region that influence quantitative intermediate risk factors for coronary artery disease," *Am. J. Hum. Genet.* 59 (1996): A29; M. B. Haviland, R. E. Ferrell, and C. F. Sing, "Association between common alleles of the low-density lipoprotein receptor gene region and interindividual variation in plasma lipid and apolipoprotein levels in a population-based sample from Rochester, Minnesota," *Hum. Genet.* 99 (1997): 108–114.

29. B. R. Murphy and P. L. Collins, "Live-attenuated virus vaccines for respiratory syncytial and parainfluenza viruses: Applications of reverse genetics," *J. Clin. Invest.* 110 (2002): 21–27.

30. K. C. Nicolaou, R. Hughes, S. Y. Cho, N. Winssinger, C. Smethurst, H. Labischinski, and R. Endermann, "Target-accelerated combinatorial synthesis and discovery of highly potent antibiotics effective against vancomycin-resistant bacteria," *Angewandte Chemie International Edition* 39 (2000): 3823–3828.

31. J. R. Cherry and A. L. Fidantsef, "Directed evolution of industrial enzymes: An update," *Curr. Opin. Biotechnol.* 14 (2003): 438–443; P. A. Dalby, "Optimising enzyme function by directed evolution," *Curr. Opin. Structural Biol.* 13 (2003): 500–505.

32. Randolph M. Nesse and George C. Williams, *Why We Get Sick: The New Science of Darwinian Medicine* (New York: Times Books, 1994); S. C. Stearns and D. Ebert, "Evolution in health and disease," *Quarterly Rev. Biol.* 76 (2001): 417–432.

33. M. D. Leakey and R. L. Hay, "Pliocene footprints in the Laetoli Beds at Laetoli, northern Tanzania," *Nature* 278 (1979): 317–323; T. D. White, G. Suwa, and B. Asfaw, "*Australopithecus ramidus,* a new species of early hominid from Aramis, Ethiopia," *Nature* 371 (1994): 306–312; Milford Wolpoff, *Human Evolution* (New York: McGraw-Hill, 1995).

34. B. I. Strassmann and R. I. M. Dunbar, "Human evolution and disease: Putting the stone age in perspective," in *Evolution in Health & Disease*, ed. S. C. Stearns (New York: Oxford University Press, 1999), pp. 91–101.

35. See S. B. Eaton and S. B. Eaton III, "The evolutionary context of chronic degenerative diseases," in *Evolution in Health & Disease,* ed. S. C. Stearns (New York: Oxford University Press, 1999), pp. 251–259; Strassmann and Dunbar, "Human evolution."

36. Mel Greaves, *Cancer: The Evolutionary Legacy* (New York: Oxford University Press, 2000).

37. G. C. Williams, "Pleiotropy, natural selection and the evolution of senescence," *Evolution* 11 (1957): 398–411; Steven N. Austad, *Why We Age* (New York: J. Wiley & Sons, 1997).

4. Evolution and Conservation

1. K. O. Stetter, "Extremophiles and their adaptation to hot environments," *FEBS Letters* 452 (1999): 22–25.
2. B. C. Christner, E. Mosley-Thompson, L. G. Thompson, and J. N. Reeve, "Isolation of bacteria and 16S rDNAs from Lake Vostok accretion ice," *Environm. Microbiol.* 3 (2001): 570–577.
3. Richard C. Lewontin, *The Triple Helix: Gene, Organism, and Environment* (Cambridge, MA: Harvard University Press, 2001).
4. C. R. Woese, "Interpreting the universal phylogenetic tree," *Proc. Nat. Acad. Sci. USA* 97 (2000): 8392–8396; E. V. Koonin, K. S. Marakova, and L. Aravind, "Horizontal gene transfer in prokaryotes: Quantification and classification," *Annu. Rev. Microbiol.* 55 (2001): 709–742; W. F. Doolittle, "Bacteria and Archaea," in *Assembling the Tree of Life,* ed. J. Cracraft and M. J. Donoghue (New York: Oxford University Press, 2004), pp. 86–94.
5. R. L. Tatusov, N. D. Fedorova, J. D. Jackson, A. R. Jacobs, B. Kiryutin, E. V. Koonin, D. M. Krylov, R. Mazumder, S. L. Mekhedov, A. N. Nikolskaya, B. S. Rao, S. Smirnov, A. V. Sverdlov, S. Vasudevan, Y. I. Wolf, J. J. Yin, and D. A. Natale, "The COG database: An updated version includes eukaryotes," *BMC Bioinformatics* 4 (2003): 41.
6. S. L. Baldauf, "The deep roots of eukaryotes," *Science* 300 (2003): 1703–1706.
7. G. M. Mace, J. E. M. Baillie, S. R. Beissinger, and K. H. Redford, "Assessment and management of species at risk," in *Conservation Biology,* ed. M. E. Soulé and G. H. Orians (Washington, DC: Soc. for Conservation Biology, Island Press, 2001), pp. 11–29; Edward O. Wilson, *The Future of Life* (New York: Alfred A. Knopf, 2002).
8. Andrew Dobson, *Conservation and Biodiversity* (New York: Scientific American Library, no. 59, W. H. Freeman and Co., 1996).
9. World Resources Institute, *World Resources 2002–2004* (New York: Oxford University Press, 2003); Richard E. Schultes and Robert F. Raffauf, *The Healing Forest: Medicinal and Toxic Plants of the Northwest Amazonia* (Portland: Dioscorides Press, 1990); John Tuxill, *Nature's Cornucopia: Our Stake in Plant Diversity* (Washington, DC: World Watch Institute, 1999).
10. Roy G. Van Driesche and Thomas S. Bellows, Jr., *Biological Control* (New York: Chapman and Hall, 1996).
11. R. Costanza, R. d'Arge, R. de Groot, S. Farber, M. Grasso, B. Hannon, K. Limburg, S. Naeem, R. V. O'Neill, J. Paruelo, R. G. Raskin, P. Sutton, and M. van den Belt, "The value of the world's ecosystem ser-

vices and natural capital," *Nature* 387 (1997): 253–260; F. Villa, M. A. Wilson, R. de Groot, S. Farber, R. Costanza, and R. M. J. Boumans, "Designing an integrated knowledge base to support ecosystem services valuation," *Ecol. Economics* 41 (2002): 445–456; National Research Council, *Valuing Ecosystem Services: Toward Better Environmental Decision-Making* (Washington, DC: The National Academies Press, 2005); see http://esd.uvm.edu for work refining this approach.

12. From a speech on 22 October 2003 by David Kemp, Australian Federal Minister for the Environment and Heritage (seen at www.deh.gov.au).

13. Jared Diamond, *Collapse: How Societies Choose to Fail or Succeed* (New York: Viking, 2005).

14. Loyal Rue, *Religion Is Not about God* (New Brunswick: Rutgers University Press, 2005), p. 366.

15. C. H. Daugherty, A. Cree, J. M. Hay, and M. B. Thompson, "Neglected taxonomy and continuing extinctions of tuatara *(Sphenodon),*" *Nature* 347 (1990): 177–179; N. J. Nelson, S. N. Keall, D. Brown, and C. H. Daugherty, "Establishing a new wild population of tuatara *(Sphenodon guntheri),*" *Conservation Biol.* 16 (2002): 887–894.

16. T. N. Engstrom, H. B. Shaffer, and W. McCord, "Phylogenetic diversity of endangered and critically endangered southeast Asian softshell turtles (Trionychidae: *Chitra*)," *Biol. Conservation* 104 (2002): 173–179; H. C. Rosenbaum, R. L. Brownell, M. W. Brown, C. Schaeff, V. Portway, B. N. White, S. Malik, L. A. Pastene, N. J. Patenaude, C. S. Baker, M. Goto, P. B. Best, P. J. Clapham, P. Hamilton, M. Moore, R. Payne, V. Rowntree, C. T. Tynan, J. L. Bannister, and R. DeSalle, "World-wide genetic differentiation of *Eubalaena:* Questioning the number of right whale species," *Mol. Ecol.* 9 (2000): 1793–1802; M. Kennedy, H. G. Spencer, "Phylogeny, biogeography, and taxonomy of Australasian teals," *The Auk* 117 (2000): 154–163.

17. A. L. Roca, N. Georgiadis, J. Pecon-Slattery, S. J. O'Brien. "Genetic evidence for two species of elephant in Africa," *Science* 293 (2001): 1473–1477; Lori S. Eggert, Caylor A. Rasner, and David S. Woodruff, "The evolution and phylogeography of the African elephant inferred from mitochondrial DNA sequence and nuclear microsatellite markers," *Proc. Roy. Soc. Lond. B* 269 (2002): 1993–2006.

18. R. M. Zink, "The role of subspecies in obscuring avian biological diversity and misleading conservation policy," *Proc. Roy. Soc. Lond. B* 271 (2004): 561–564; J. A. Johnson, R. T. Watson, and D. P. Mindell, "Prioritizing species conservation: Does the Cape Verde kite exist?" *Proc. Roy. Soc. Lond. B* 272 (2005): 1365–1371.

19. R. K. Wayne, "Conservation genetics in the Canidae," in *Conservation Genetics: Case Histories from Nature,* ed. J. C. Avise and J. L. Hamrick (New York: Chapman & Hall, 1996), pp. 75–118 ; P. J. Wilson, S. Grewal, T. McFadden, R. C. Chambers, and B. N. White, "Mitochondrial DNA extracted from eastern North American wolves killed in the 1800s is not of gray wolf origin," *Canadian J. Zool.* 81 (2003): 936–940.
20. J. M. Rhymer and D. Simberloff, "Extinction by hybridization and introgression," *Annu. Rev. Ecol. Syst.* 27 (1996): 83–109.
21. F. W. Allendorf, R. F. Leary, P. Spruell, and J. K. Wenburg, "The problems with hybrids: Setting conservation guidelines," *Trends Ecol. Evol.* 16 (2001): 613–622.
22. T. B. Smith, M. W. Bruford, and R. K. Wayne, "The preservation of process: The missing element of conservation programs," *Biodiversity Letters* 1 (1993): 164–167; B. W. Bowen, "Preserving genes, species, or ecosystems? Healing the fractured foundations of conservation policy," *Mol. Ecol.* 8 (1999): S5–S10 Suppl. 1; C. Moritz, "Strategies to protect biological diversity and the evolutionary processes that sustain it," *Syst. Biol.* 51 (2002): 238–254.
23. M. J. Benton, *The Fossil Record 2* (New York: Chapman & Hall, 1993); J. R. Rest, J. C. Ast, C. C. Austin, P. J. Waddell, E. A. Tibbetts, J. M. Hay, and D. P. Mindell, "Molecular systematics of Reptilia and the tuatara mitochondrial genome," *Mol. Phylogenet. Evol.* 29 (2003): 289–297.
24. M. Lynch, "A quantitative-genetic perspective on conservation issues," in *Conservation Genetics: Case Histories from Nature,* ed. J. C. Avise and J. L. Hamrick (New York: Chapman & Hall, 1996), pp. 471–501.
25. T. L. Erwin, "An evolutionary basis for conservation strategies," *Science* 253 (1991): 750–752.
26. L. Hanna, "Calanolide A: A natural non-nucleoside reverse transcriptase inhibitor," *BETA* 12 (1999): 8–9.
27. C. S. Baker and S. R. Palumbi, "Which whales are hunted? Molecular genetic evidence for illegal whaling," *Science* 265 (1994): 1538–1539; A. Dizon, G. Lento, S. Baker, P. Parsboll, F. Capriano, and R. Reeves, *Molecular Genetic Identification of Whales, Dolphins, and Porpoises: Proceedings of a Workshop on the Forensic Use of Molecular Techniques to Identify Wildlife Products in the Marketplace* (Washington, DC: NOAA Technical Memorandum NMFS. US Department of Commerce, 2000).
28. T. L. Goldberg, "Inferring the geographic origins of 'refugee' chim-

panzees in Uganda from mitochondrial DNA sequences." *Conserv. Biol.* 11 (1997): 1441–1446.

29. For a general treatment, see Richard Frankham, Jonathan D. Ballou, and David A. Briscoe, *Introduction to Conservation Genetics* (Cambridge: Cambridge University Press, 2002).

30. For a test and illustration of the coevolutionary and adaptive nature of hummingbird bills and their associated flowering food plants, see E. J. Temeles and W. J. Kress, "Adaptation in a plant-hummingbird association," *Science* 300 (2003): 630–633.

31. P. A. Cox, T. Elmqvist, E. D. Pierson, and W. E. Rainey, "Flying foxes as strong interactors in South-Pacific island ecosystems—A conservation hypothesis," *Conserv. Biol.* 5 (1991): 448–454; S. Banack, "Diet selection and resource use by flying foxes (genus *Pteropus*)," *Ecology* 79 (1998): 1949–1967.

32. J. Estes and J. Palmisano, "Sea otters: Their role in structuring nearshore communities," *Science* 185 (1974): 1058–1060; J. A. Estes, M. T. Tinker, T. M. Williams, and D. F. Doak, "Killer whale predation on sea otters linking oceanic and nearshore ecosystems," *Science* 282 (1998): 473–476.

33. R. S. Ostfeld and F. Keesing, "Biodiversity and disease risk: The case of Lyme disease," *Conserv. Biol.* 14 (2000): 722–728.

34. F. S. Chapin, O. E. Sala, I. C. Burke, J. P. Grime, D. U. Hooper, W. K. Lauenroth, A. Lombard, H. A. Mooney, A. R. Mosier, S. Naeem, S. W. Pacala, J. Roy, W. L. Steffen, and D. Tilman, "Ecosystem consequences of changing biodiversity—Experimental evidence and a research agenda for the future," *BioScience* 48 (1998): 45–52.

5. Evolutionary Metaphor in Human Culture

1. D. B. Searls, "The linguistics of DNA," *Am. Scient.* 80 (1992): 579–591; D. B. Searls, "From Jabberwocky to genome: Lewis Carroll and computational biology," *J. Computational Biol.* 8 (2001): 339–348; D. B. Searls, "Trees of life and of language," *Nature* 426 (2003): 391–392.

2. Robert M. W. Dixon, *The Rise and Fall of Languages* (Cambridge: Cambridge University Press, 1997).

3. M. Swadesh, "Lexico-statistic dating of prehistoric ethnic contacts," *Proc. Am. Philosoph. Soc.* 96 (1952): 452–463.

4. M. Pagel, "Maximum-likelihood models for glottochronology and for reconstructing linguistic phylogenies," in *Time Depth in Historical Linguistics, Vol. 1,* ed. C. Renfrew, A. McMahon, and L. Trask (Oxford: The McDonald Inst. for Archaeological Research, Oxbow Books,

2000), pp. 189–222; R. D. Gray and Q. D. Atkinson, "Language-tree divergence times support the Anatolian theory of Indo-European origin," *Nature* 426 (2003): 435–439; Q. D. Atkinson and R. D. Gray, "Curious parallels and curious connections—phylogenetic thinking in biology and historical linguistics," *Syst. Biol.* 54 (2005): 513–526.

5. Johanna Nichols, *Linguistic Diversity in Time and Space* (Chicago: University of Chicago Press, 1992).
6. J. Diamond and P. Bellwood, "Farmers and their languages: The first expansions," *Science* 300 (2003): 597–603.
7. Sources consulted on the history of biblical translations and used in making Figure 5.4 are: Alister McGrath, *In the Beginning* (New York: Anchor Books, 2001); Melvin Konner, *Unsettled: An Anthropology of the Jews* (New York: Viking Compass, 2003); Jaroslav Pelikan, *Whose Bible Is It? A History of the Scriptures through the Ages* (New York: Viking, 2005); and Stanley L. Greenslade, ed., *Cambridge History of the Bible* (Cambridge: Cambridge University Press, 1975).
8. Alister McGrath, *In the Beginning* (New York: Anchor Books, 2001).
9. Sources for cultural evolution of Judaism, Christianity, and Islam used for text discussion and in making Figures 5.5–5.8 include: Mircea Eliade, ed., *The Encyclopedia of Religion* (New York: Macmillan, 1987); Konner, *Unsettled;* Paul Johnson, *A History of the Jews* (New York: Harper & Row, 1987); P. M. Holt, Ann K. S. Lambton, and Bernard Lewis, eds., *The Cambridge History of Islam* (Cambridge: Cambridge University Press, 2000); Eliot Shaw and Michael Pye, eds., *Overview of World Religions* (philtar.ucsm.ac.uk/encyclopedia; accessed 20 March 2005).
10. Konner, *Unsettled.*
11. Matt Ridley, *The Origins of Virtue* (New York: Penguin Books, 1996), p. 191.
12. Richard D. Alexander, *The Biology of Moral Systems* (New York: Aldine de Gruyter, Hawthorne, 1987).
13. Alexander Heidel, *The Gilgamesh Epic and Old Testament Parallels* (Chicago: University of Chicago Press, 1949); Werner Keller, *The Bible as History* (New York: William Morrow and Company, 1956); Nancy K. Sandars, *The Epic of Gilgamesh (An English Translation with Introduction)* (London: Penguin Books, 1964).
14. Richard C. Foltz, *Religions of the Silk Road* (New York: St. Martin's Press, 1999).
15. Tudor Parfitt, *The Lost Tribes of Israel* (London: Orion Books, 2002).
16. K. Skorecki, S. Selig, S. Blazer, R. Bradman, N. Bradman, P. J. Waburton, M. Ismajlowicz, and M. F. Hammer, "Y chromosomes of Jewish

priests," *Nature* 385 (1997): 32; M. G. Thomas, K. Skorecki, M. Haim Ben-Amid, T. Parfitt, N. Bradman, and D. B. Goldstein, "Origins of Old Testament priests," *Nature* 394 (1998): 138–140.

17. C. Piazzi Smyth, *Our Inheritance in the Great Pyramid* (London: W. Isbister, 1864).

18. Doctrine and Covenants of the Church of Jesus Christ of Latter-Day Saints 133:26, quotation in Parfitt, *Lost Tribes,* p. 110.

19. R. N. Hall, *Great Zimbabwe* (London: Methuen, 1902), quoted in Parfitt, *Lost Tribes.*

20. M. G. Thomas, T. Parfitt, D. A. Weiss, K. Skorecki, J. F. Wilson, M. le Roux, N. Bradman, and D. B. Goldstein, "Lemba and the Cohen Modal Haplotype," *Am. J. Hum. Genet.* 66 (2000): 674–686; Thomas et al., "Origins."

21. For example, Antonio Torroni, "Mitochondrial DNA and the origin of Native Americans," in *America Past, America Present: Genes and Languages in the Americas and Beyond,* ed. Colin Renfrew (Cambridge: McDonald Institute for Archaeological Research, 2000).

22. E. S. Poloni, O. Semino, G. Passarino, A. S. Santachiara-Benerecetti, I. Dupanloup, A. Langaney, and L. Excoffier, "Human genetic affinities for Y-chromosome P49a,f/TaqI haplotypes show strong correspondence with linguistics," *Am. J. Hum. Genetics* 61 (1997): 1015–1035; T. M. Karafet, S. L. Zegura, O. Posukh, L. Osipova, A. Bergen, J. Long, D. Goldman, W. Klitz, S. Harihara, P. de Knijff, V. Wiebe, R. C. Griffiths, A. R. Templeton, and M. F. Hammer, "Ancestral Asian source(s) of New World Y-chromosome founder haplotypes," *Am. J. Hum. Genetics* 64 (1999): 817–831; L. L. Cavalli-Sforza and M. W. Feldman, "The application of molecular genetic approaches to the study of human evolution," *Nature Genetics* 33 (2003): 266–275; P. A. Underhill, "Inferring human history: Clues from Y-chromosome haplotypes," *Cold Spring Harbor Symposia on Quantitative Biology* LXVII (2003): 487–493.

23. Luigi Luca Cavalli-Sforza, *Genes, Peoples, and Languages* (Berkeley: University of California Press, 2000).

6. The Role of Evolution in Court and Classroom

1. Edward J. Larson, *Trial and Error,* 3rd ed. (Oxford: Oxford University Press, 2003).

2. For more information on the applications and distinctions of the *Frye* and *Daubert* rulings to forensics, see R. Harmon, "Admissibility standards for scientific evidence," Roger G. Breeze, Bruce Budowle, and

Steven E. Schutzer, eds., *Microbial Forensics* (Burlington, MA: Elsevier Academic Press, 2005), 381–392.

3. A. J. Jeffreys, V. Wilson, and S. L. Thein, "Hypervariable 'minisatellite' regions in human DNA," *Nature* 314 (1985): 67–73.

4. See most recent figures compiled by The Innocence Project at www.innocenceproject.org.

5. M. L. Metzker, D. P. Mindell, X.-M. Liu, R. G. Ptak, R. A. Gibbs, and D. M. Hillis, "Molecular evidence of HIV-1 transmission in a U.S. criminal case," *Proc. Nat. Acad. Sci. USA* 99 (2002): 14292–14297.

6. J. Huelsenbeck and F. Ronquist, "MRBAYES: Bayesian inference of phylogenetic trees," *Bioinformatics* 17 (2001): 754–755.

7. J. Albert, J. Wahlberg, T. Leitner, D. Escanilla, and M. Uhlen, "Analysis of a rape case by direct sequencing of the human-immunodeficiency-virus type-1 *pol* and *gag* genes," *J. Virol.* 68 (1994): 5918–5924; R. Machuca, L. B. Jorgensen, P. Theilade, and C. Nielsen, "Molecular investigation of transmission of human immunodeficiency virus type 1 in a criminal case," *Clinical and Diagnostic Lab. Immunol.* 8 (2001): 884–890; C. Y. Ou, C. A. Ciesielski, G. Myers, C. I. Bandea, C. C. Luo, B. T. M. Korber, J. I. Mullins, G. Schochetman, R. L. Berkelman, A. N. Economou, J. J. Witte, L. J. Furman, G. A. Satten, K. A. Macinnes, J. W. Curran, and H. W. Jaffe, "Molecular epidemiology of HIV transmission in a dental practice," *Science* 256 (1992): 1165–1171.

8. M. Lee Goff, *A Fly for the Prosecution* (Cambridge, MA: Harvard University Press, 2000).

9. See web pages on forensic entomology by J. H. Byrd at www.forensic-entomology.com.

10. From Wayne D. Lord, "Case Histories of the use of insects in investigations," Federal Bureau of Investigation, Washington, DC; www.missouri.edu/~agwww/entomology/casestudies.

11. G. E. Allen, "Eugenics," in *Encyclopedia of Life Sciences* (London: Nature Publishing Group, 2001); www.els.net.

12. For details on Galton see the excellent biography by Nicholas W. Gilham, *A Life of Sir Francis Galton: From African Exploration to the Birth of Eugenics* (Oxford: Oxford University Press, 2001).

13. S. Ribeiro and G. B. Golding, "The mosaic nature of the eukaryotic nucleus," *Mol. Biol. Evol.* 15 (1998): 779–788; C. R. Woese, "On the evolution of cells," *Proc. Nat. Acad. Sci. USA* 99 (2002): 8742–8747; W. F. Doolittle, "Bacteria and Archaea," in *Assembling the Tree of Life,* J. Cracraft and M. J. Donoghue (New York: Oxford University Press, 2004), pp. 86–94; N. R. Pace, "The early branches in the tree of life,"

in *Assembling the Tree of Life,* ed. J. Cracraft and M. J. Donoghue (New York: Oxford University Press, 2004), pp. 76–85.

14. William D. Hamilton, *Narrow Roads of Gene Land, vol. 2* (New York: W.H. Freeman/Spektrum, 2002).
15. C. M. Lively, "Evidence from a New Zealand snail for the maintenance of sex by parasitism," *Nature* 328 (1987): 519–521; C. M. Lively and M. F. Dybdahl, "Parasite adaptation to locally common host genotypes," *Nature* 405 (2000): 679–681.
16. S. A. West, C. M. Lively, and A. F. Read, "A pluralist approach to the evolution of sex and recombination," *J. Evol. Biol.* 12 (1999): 1002–1012; Hamilton, *Narrow Roads.*
17. E. Williams, W. Crossley, and T. Lang, "Average and maximum revisit time trade studies for satellite constellations using a multiobjective genetic algorithm," *J. Astronautical Sci.* 49 (2001): 385–400.
18. J. E. Beasley, J. Sonander, and P. Havelock, "Scheduling aircraft landings at London Heathrow using a population heuristic," *J. Operational Res. Soc.* 52 (2001): 483–493.
19. Philip E. Johnson, *Darwin on Trial* (Washington, DC: InterVarsity Press, 1991); Michael J. Behe, *Darwin's Black Box* (New York: Touchstone, 1996); William A. Dembski, *The Design Inference: Eliminating Chance through Small Probabilities* (Cambridge: Cambridge University Press, 1998); "Evolution Wars," *Time,* August 15, 2005, p. 27.
20. For discussion and refutation of "intelligent design" claims see: Robert T. Pennock, ed., *Intelligent Design Creationism and Its Critics: Philosophical, Theological, and Scientific Perspectives* (Boston: The MIT Press, 2001); H. Allen Orr, "Book review: No free lunch," *Boston Review: A Political and Literary Forum* 27 (2002): nos. 3–4; Kenneth R. Miller, "The flagellum unspun: The collapse of 'irreducible complexity,'" in *Debating Design: From Darwin to DNA,* ed. M. Ruse and W. Dembski (Cambridge: Cambridge University Press, 2004); Mark Perakh, *Unintelligent Design* (Amherst, NY: Prometheus Books, 2004).
21. Phillip E. Johnson, "Enlisting science to find the fingerprints of a creator," *LA Times,* 25 March 2001.
22. Phillip E. Johnson, American Family Radio, 10 Jan. 2003, quoted from the Talk.Origins web pages at www.talkorigins.org.
23. Stephen Jay Gould, *Rocks of Ages: Science and Religion in the Fullness of Life* (New York: Ballantine, 1999).
24. Richard D. Alexander, *The Biology of Moral Systems* (New York: Aldine de Gruyter, Hawthorne, 1987); W. B. Provine, "Scientists, face it! Science and religion are incompatible," *The Scientist,* 5 September 1988, p. 10; Matt Ridley, *The Origins of Virtue* (New York: Penguin Books,

1996); Kenneth R. Miller, *Finding Darwin's God* (New York: Harper Collins, 1999); Massimo Pigliucci, *Tales of the Rational* (Atlanta: Freethought Press, 2000); Richard Dawkins, *A Devil's Chaplain* (Boston: Houghton Mifflin, 2003).

Conclusions

1. Walter Isaacson, *Benjamin Franklin: An American Life* (New York: Simon & Schuster, 2003).
2. For further information on applications and teaching of evolutionary biology, see Joel Cracraft and Rodger W. Bybee, eds., *Evolutionary Science and Society: Educating a New Generation* (Colorado Springs, CO: BSCS, 2005).
3. Phillip E. Johnson, *Reason in the Balance: The Case Against Naturalism in Science, Law and Education* (Downers Grove, IL: InterVarsity Press, 1995).
4. D. M. Hillis, "The tree of life and the grand synthesis of biology," in *Assembling the Tree of Life,* ed. J. Cracraft and M. J. Donoghue (New York: Oxford University Press, 2004), pp. 545–547.
5. From a taped lecture given in the Fellowship Baptist Church in Waco, Texas, 7 March 2004, quoted in Mark Perakh, "The Design Revolution?" online at www.talkdesign.org/people/mperakh/perakh_ddq.pdf, p. 5.
6. James D. Hunter, *Culture Wars* (New York: Basic Books, 1991), p. 52; Kary D. Smout, *The Creation/Evolution Controversy: A Battle for Cultural Power* (Westport, CT: Praeger, 1998), p. 184.
7. Ian G. Barbour, *When Science Meets Religion* (New York: Harper Collins, 2000).
8. Gallup News Service, press release, 5 March 2001.
9. Loyal Rue, *Religion Is Not about God* (New Brunswick: Rutgers University Press, 2005).
10. Jacob Bronowski, *Science and Human Values* (New York: Harper & Row, 1956).

ILLUSTRATION CREDITS

Original drawings on p. ii and in Chapters 2 and 3 are the work of John Megahan.

Figure 3.7 Redrawn from P. A. Dalby, "Optimising enzyme function by directed evolution," *Curr. Opin. Structural Biol.* 13 (2003): 500–505.

Figure 4.2 Based on M. Kennedy and H. G. Spencer, "Phylogeny, biogeography, and taxonomy of Australasian teals," *The Auk* 117 (2000): 154–163; A. L. Roca, N. Georgiadis, J. Pecon-Slattery, and S. J. O'Brien, "Genetic Evidence for Two Species of Elephant in Africa," *Science* 293 (2001): 1473–1477; L. S. Eggert, C. A. Rasner, and D. S. Woodruff, "The evolution and phylogeography of the African elephant inferred from mitochondrial DNA sequence and nuclear microsatellite markers," *Proc. Roy. Soc. Lond. B* 269 (2002): 1993–2006; and J. A. Johnson, R. T. Watson, and D. P. Mindell, "Prioritizing species conservation: Does the Cape Verde kite exist?" *Proc. Roy. Soc. Lond. B* 272 (2005): 1365–1371.

Figure 4.3 Redrawn from B. W. Bowen, "Preserving genes, species, or ecosystems? Healing the fractured foundations of conservation policy," *Mol. Ecol.* 8 (1999): S5-S10 Suppl. 1.

Figure 4.4 Redrawn and expanded from C. S. Baker and S. R. Palumbi, "Which whales are hunted? Molecular genetic evidence for illegal whaling," *Science* 265 (1994): 1538–1539, and A. Dizon, G. Lento, S. Baker, P. Parsboll, F. Capriano, and R. Reeves, *Molecular Genetic Identification of Whales, Dolphins, and Porpoises: Proceedings of a Workshop on the Forensic Use of Molecular Techniques to Identify Wildlife Products in the Marketplace* (Washington, DC: NOAA Technical Memorandum NMFS. US Department of Commerce, 2000).

Figure 5.1 Based on D. B. Searls, "From Jabberwocky to genome: Lewis Carroll and computational biology," *J. Computational Biol.* 8 (2001): 339–348; M. Gardner, *The Universe in a Handkerchief: Lewis Carroll's Mathematical Recreations, Games, Puzzles, and Word Plays* (New York: Copernicus/Springer-Verlag, 1996); and E. Wakeling, *Lewis Carroll's Games and Puzzles* (New York: Dover, 1992).

Figure 5.2 After R. D. Gray and Q. D. Atkinson, "Language-tree divergence times support the Anatolian theory of Indo-European origin," *Nature* 426 (2003): 435–439.

Figure 5.3 After D. B. Searls, "Trees of life and of language," *Nature* 426 (2003): 391–392.

Figure 5.9 After L. L. Cavalli-Sforza and M. W. Feldman, "The application of molecular genetic approaches to the study of human evolution," *Nature Genetics* 33 (2003): 266–275.

INDEX